Gibbons in the Family Tree

Gibbons in the Family Tree

Jeanne Ann Vanderhoef

Brunswick

Copyright © 1996 by Jeanne Ann Vanderhoef

All rights reserved. No part of this book may be reproduced in any form or by any means, electronic or mechanical, including photocopying or by any informational storage or retrieval system, without written permission from the author and the publisher.

Library of Congress Cataloging-in-Publication Data

Vanderhoef, Jeanne Ann, 1918–
    Gibbons in the family tree / Jeanne Ann Vanderhoef. — 1st ed.
      p.   cm.
    ISBN 1-55618-161-2 (hardcover : alk. paper)
    1. Gibbons as pets—Anecdotes.   2. Gibbons—Virginia—Anecdotes.   3. Vanderhoef, Jeanne Ann, 1918– .   I. Title.
SF459.G53V36   1996
636'.9882—dc20
[B]   96-35122
                                                                CIP

First Edition

Published in the United States of America

by

Brunswick Publishing Corporation

1386 Lawrenceville Plank Road
Lawrenceville, Virginia 23868
1-800-336-7154

For Van, Craig, Lee and Christy, the love and strength of the family tree, . . . for Janey and all my other friends who either laughed or cried over the chapters as they were born . . . and most of all for the furry members of our family tree who filled our branches, lives and hearts to overflowing!

I

I was on my way to buy a gibbon! I had been waiting two months for this day to arrive, ever since our second day in Bangkok when I had first seen one of the little apes. Van, my husband, had promised that as soon as we were settled in a house I could have one, and now, with my fourteen-year-old son, Craig, and our driver, Bunchu, I was going to the pet shop.

"I'll sit in front with Bunchu," Craig said as he slid onto the hot leather seat of the steamy car.

Already in the back seat, I nodded and braced myself for the jerk I knew was coming.

Bunchu was waiting in gear with one foot on the brake and the other on the clutch. As Craig's door slammed, Bunchu removed both feet at once and we catapulted through the open gates. If he'd known the words, he'd have yelled, "Varoom, varoom," as he tramped on the accelerator.

Round-faced Bunchu was in his twenties and loved to drive. It was a skill that must have been recently acquired. He had two speeds forward—that of a torpedo with an unswerving trajectory, and the other—slower than a clogged catsup bottle. The latter was dictated by the traffic on roads designed for nothing larger than samlors (pedicabs) and bicycles. According to Bunchu, the inevitable grid-lock was a plot against Thailand by the Japanese who had occupied the country during World War II.

Whatever the reason, in 1956, the roads were constricted by cars, motor scooters, pedicabs, bicycles and trucks. Once in a while a few water buffalo were herded down the main thoroughfares, but the trucks felt they had the right of way and the herder who valued his own life as well as those of his buffalo, usually chose a more circuitous and safer route.

After our explosion from the gate, the driver draped his right arm across the back of the front seat and we shot onto Pattipat (pronounced Pottypot) Road. The traffic, for a moment, was light, and we careened down the narrow strip, bouncing from pothole to pothole and skidded across a narrow bridge at the end. As we swished around the corner onto Don Muang Road, I could see that the vehicles a quarter of a mile ahead were beginning to form a clot in the city's artery. Bunchu ignored the signs and we sped on.

Coming toward us was a bus with passengers clinging to the top and sides; ahead of us was a huge charcoal truck, fully loaded, with four men and assorted baskets perched on top of the grimy briquettes. The grinning men waved us on to pass them and Bunchu accepted the challenge.

They were abreast of each other, the truck and the bus, when we, like a fat arrow, flew between them.

My eyes were closed and I neither saw nor felt the car suck in its sides, but it must have because we emerged on the other side alive and with no Thai bodies splattered on our hood.

"Oh, Bunchu," I gasped when I could exhale.

Our driver took his eyes off the road, turned in his seat and gave me a wide grin. He languidly raised the arm that still lay limply along the seat back, flapped his hand at me and said soothingly, "Never mind, Madame." It was his stock answer to everything upsetting and, in view of the fact that I was going to be riding with him for the next two years, I thought it a philosophy I would do well to accept.

Van was a colonel in the United States Army and, as the American Advisor to the Thai Intelligence School, was given two cars. He drove the small Borg Ward back and forth to his office himself, but the car that the family used, the one driven by Bunchu, was a four-door Dodge with leather seats and no air-conditioning to alleviate the tropical climate. Because the Dodge was bigger and had more power, Bunchu was satisfied with the arrangements.

His hours were long, starting when he took the children to the International School in the morning and ending when he delivered Van and me to and from whatever evening function we had to attend; but he napped a lot so I didn't worry about him.

The charcoal truck and bus spat us out into the traffic jam and the noise was unbelievable. Vespa motor scooters beeped, taxis and cars cursed each other with their horns, and public address systems blared songs like "Bernadine," sung in high sing-songy voices, from each open-fronted store along the road side. Ahead of us, a ramshackle truck carrying two enormous dead shark filled the car with carbon monoxide and putrefying fish aroma.

Forty-five minutes later, still behind the further-aging shark, we inched along the dusty road. I had the feeling we were blocked by a solid curtain of heavy, smelly heat that was actively pressing us backward. The puddle of perspiration that dripped from the heels of my open sandals onto the rubber floor mat, grew in depth as I tried to think about my gibbon-to-be.

We had been invited to a luncheon on our second day in

Bangkok and the hosts had a young gibbon who lived in their garden. He was attached by a belt and chain to a wire that ran between two trees and, as I approached him, he opened his arms and ran toward me on his hind legs with a happy grin. When I picked him up and his long arms wrapped around my neck, I knew I had to have a furry baby of my own. Now the day that Van had promised me had arrived and I resented every car on the road that kept me from getting my gibbon.

"Madame, Claig, go walk." Bunchu's voice startled me. "See-look shop. Not smell bad so much."

Craig was already opening his door into the stalled traffic. "C'mon, Mom, let's go. We'll be cooler walking and we'll be away from those shark."

"All right." I stepped out into the road. "Where will you meet us, Bunchu?" I slammed the door and waited.

"Near-by Emlard Buddha," he said. I knew where that was and dodged away between the vehicles.

On the sidewalk a huge fish gazed at me with glassy eyes. It basked in the searing sun on a banana leaf laid on the pavement, and I realized that we were at the open market, the one that provided us with most of our food. Pi, the cook, did the purchasing so I never before had been this close to the source of our meats and vegetables. I was immediately sorry that the opportunity had now presented itself.

"Craig, hurry up." My tummy lurched and I put my hand over my mouth and nose. "Let's get by here quickly. If we don't, I may never eat again!"

Craig stared at the food spread on the banana leaf plates arranged on the sidewalk. A vendor grinned at him with red, beetle-nut stained gums and few teeth. He shooed the flies away with a wispy broom. They circled, buzzed and relit at once. Craig shuddered.

"I'd just as soon forget the bare feet that have tramped over my future dinner, myself," he said as we pushed our way through the perspiring crowds and cloying smells, "but did you see the beetle-nut chewer spit right next to the papaya?"

"I only hope I can forget it!" I had seen the woman spit and had noticed the tell-tale red stains in the dust near almost every leaf on the concrete. "Thank God for Clorox." I referred to the fact that all the vegetables or fruits we used that weren't cooked were soaked in straight chlorine bleach to kill the flukes, amoebas and, we hoped, bacteria.

Something brushed my hair and I turned my face into a dead duck that looked as though it had been flattened by a steam-roller. Its head hung limply on my shoulder and it stared at me with squashed, lifeless eyes. A whole row of them hung from a line across the open storefront, and I wondered how the Thai housewives in their colorful patungs, could tell any difference as they fingered the paper-doll thin bodies. The women also pinched and poked the naked chickens that hung by scrawny necks from a bamboo pole. The odor of the fowl, mingling with the pervasive smell of garlic, seemed to cling to everything in the oppressive heat.

We hurried past the market and turned a corner onto a quiet street. It was bordered on one side by a klong (canal) that was filled with splashing, naked boys about seven or eight years old.

"That's the life!" Craig sounded envious of the group. I understood. I wanted to jump into the klong myself. Two very little girls wearing only aprons that stopped short of naked, brown thighs, skipped ahead of us. Suddenly, they crossed the road and jumped into the canal with the boys. Now I was envious.

"Look at them," Craig continued. "They can cool themselves in the klongs, pick a banana or papaya when they're hungry and haven't a care in the world."

"You've got to be kidding," I said as I thought of the perils these kids faced daily. "What about the poisonous snakes in every bush and tree, the leeches in the water, rabies, T.B., to say nothing of the flukes and amoebic dysentery? Those seem like pretty big worries to me."

"Kids don't worry about those things." Craig gave me an innocent look and shrugged his shoulders. "That's a job for their mothers."

He was right, of course but I just shook my head. I wouldn't rise to the bait.

A pack of wild dogs, mangy skin hanging in festoons from emaciated bodies, snarled weakly at each other as they foraged for scraps in the gutter. None appeared rabid, but one, an obvious mother, was so thin I wondered how she could stand up or what she had to feed the puppies. I hoped she didn't have many to add to the canine population. My heart wept for her and for the thousands that roamed the streets, unfed and unloved.

"You know, Craig," I sighed, "this country has so many interesting things. The people are beautiful and friendly. They're never cold and never hungry. They have flowers and antiquities and delicious, exotic fruits but—as an animal lover—I'm afraid my time here will be colored by things like these pitiful dogs. The Thai seem so cruel and callous as far as animals are concerned."

"Bunchu says their religion teaches that a dog is a reincarnation of an evil person, therefore to be punished."

Craig spent a lot of time with Bunchu. "It's hard to believe, but I'm trying to learn not to see the dogs."

"Any success?" I asked knowing what the answer would be. None of us would be successful and would suffer for the next two years.

"Not yet, but I haven't given up." Craig was an optimist.

As we approached the end of the street, a shiny roof lacquered-red and bordered in a wide band of emerald green, rose above the trees. Along its edge, flame shapes covered in gold leaf reached toward the sun and reflected its rays back into our eyes. It was the Temple of the Emerald Buddha.

As we walked closer, the wall around the temple and its surrounding buildings cut off our view of the roof, but I knew that inside, protected from the sun, the small green Buddha smiled serenely.

Bunchu was leaning against the wall under a tree that shaded both him and the car. "No more tlaffic now," he announced happily. "We go rickerty sprit." To the Oriental ear, l's and r's are interchangeable, but I knew what Bunchu meant, no matter what consonant he used. My heart sank.

In the car I braced for the take-off. Bunchu grasped the wheel, raised his feet, slammed the right one onto the accelerator and we bounded away from the curb. We raced down the road and screeched to a halt at the klong that separated us from the pet shop. I hardly had time to get nervous.

A rickety bridge spanned the canal, and Craig and I stopped in the middle to observe, in awe, the activities going on below us.

A young woman in a black sarong stood waist deep in

the water as she washed her hair. Her dirty suds, however, swirled down the sluggish stream to surround a man in white boxer shorts who was brushing his teeth. Upstream of them both, a child attending to other bathroom functions, squatted over the edge of a plank that jutted out like a small pier into the gray-green water. A dead pig, bloated and pinky-white, floated beneath the child and toward the ablutionists through a tangle of water hyacinths.

"An old fashioned hand pump and an outhouse would be an improvement over this," Craig remarked quietly.

"Anything would be an improvement over this," I answered. "Anything!"

As Bunchu joined us on the bridge, I wondered, briefly, what the bathroom situation was in his home. He always appeared clean and starched, but then, in our two months in Thailand, I had never met a person who looked dirty or who had an unpleasant odor. Given the heat and primitive facilities in many homes, I wondered how they managed.

Across the rotting boards, the pet shop looked deceptively picturesque. Under a sagging and rusted corrugated tin roof, it seemed shady and cool. Open on the front, it was ringed around with banana and mimosa trees. Dusty hibiscus bushes in full bloom poked their branches in and around the cages of brightly colored birds that were out in front. Ducks and geese attached by long strings to trees and bushes squatted or waddled on the bank of the canal.

Under the roof, it was a different story. There, if there were any, the furry animals lived or died. Birds and other fowl were always saleable—the Thai loved them—but other pathetic creatures were kept mainly for foreigners. They were acquired cheaply by the owner and so were a small loss if they died. They stayed, often without food and water, under the steaming roof. The dilapidated cages were stacked four high in rows that formed passageways so narrow that even the fetid air could not escape.

"I guess we're out of luck," Craig said as we squeezed up and down the aisles. "I don't even see a rat."

"We have enough of those without buying one," I replied, "and your sisters have managed to revive all the guinea pigs we're ever likely to need, so I hope we don't see any more of those."

Two of the little brown and white tail-less creatures had been nearly dead when Lee and Christy had brought them home. Now they were healthy and, we were happy to find, both of the same sex.

Craig, himself, had rescued the two mongooses, Pat and Mike. Although they were still wild and had to be caged, from time to time they escaped and before they were again incarcerated, they would rack up an imposing list of cobras and other poisonous snakes. A phobia against all legless reptiles made me yearn for the day when the mongooses could be allowed to roam freely in the compound.

But the cages for the fur-bearing animals were empty as Craig and I searched for a gibbon—until the last cage of the last row.

As I peered into the gloom, two terrified eyes, round and black, looked back at me. They belonged to a very young, female gibbon. Huddled in the farthest corner of her crate, her long arms wrapped tightly around her knees, she stared at me with such fear that I stepped back as far as I could so I wouldn't frighten her further.

Her fur was matted and discolored by the filth in her cage. Only the white fringe that surrounded her tiny, pansy-shaped, black face was clean and seemed to glow in that horrid place.

I turned to the pet shop owner who, with Bunchu, was behind us.

"Tao Rai (how much)?" I asked. I hadn't yet learned much of the difficult language, but I could ask the price of an item.

"Ha loy Baht (five hundred Baht)," he answered.

As I took out my wallet, Bunchu flew into a frenzy.

"Too much cost! Too much cost! Madame not pay!" He fluttered his hands in agitation.

"Let Bunchu bargain for you, Mother," Craig said. "Otherwise, you'll lose face."

"I don't care if I do lose face." I continued to take out the money. "I won't haggle for this bedraggled bit of life as if she had no value, anymore than I would haggle for your life. She has a value to me and I'd pay any amount to get her out of here. You don't have to tell the owner that." I glared at Bunchu as I handed the man five one-hundred Baht notes, each worth five dollars.

Penny, as the gibbon was to be named, was miserable in the cage, but was in abject terror of us. I reached into the dim recesses and as soon as I touched her, she bit me. Her baby teeth barely pinched and as I got both hands around her middle, she kept nipping and making forlorn little whooping noises. Her two hands and feet worked like four hands and as I pried the fingers of one from the cage wires, another hand and two feet still clung. We had a tussle, but at last I won, and had her dangling from the end of my outstretched arm.

I had been told that adult gibbons had long, sharp eye teeth and could be dangerous. In order to get a baby for resale, the hunters went out into the jungle, shot a mother and snatched the little one. No wonder this baby was so afraid of humans.

The so-called pet shop.

As the chanee (Thai word for gibbon) turned and twisted in the air, Craig fastened a small cat collar around her waist and attached a leash so we wouldn't lose her. Bunchu, still sulking, stalked ahead of us as we wrestled our way across the bridge and into the car.

"What's she doing?" Craig asked from the front seat when we were underway.

"She's hunched in the farthest corner on the floor," I answered. "She's staring at me and, I think, trying to figure out how she can get away. Poor little thing! I wish she knew how safe she is with us and how we'll spoil her if she'll let us."

"We've had frightened things before, Mom." Craig craned his neck over the back of the seat and stared down on the chanee. She drew into an even smaller ball and stared back at him without blinking. "She'll learn," he said as he sat back down. I knew he was right.

Gibbons are the smallest and, I think, the most beautiful of the ape family which consists of gibbon, chimpanzee, orangutan and gorilla in ascending order of size. The gibbon is the only primate, aside from man, that walks entirely on hind legs, but it has long arms meant for swinging through trees.

Penny, when clean, turned out to be champagne colored. Now, as she huddled on the car floor, she was filthy from her ordeal, but I could see that she had lovely long fingers on her tiny hands and that they ended in perfectly shaped nails. Her palms were black, as were the soles of her feet, her little bottom and her white-rimmed, pansy-shaped face. Her ears, too, were black, shaped just like a human's and flat to the sides of her round, furry head.

I wanted so badly to pick her up and hug her to me as I had hugged the gibbon two months before. I wanted to feel her arms around my neck and to know that she felt secure with us and that we could exchange love . . . but I had to wait.

"Little gibbon," I said to the apprehensive baby crouched in the corner, "you're safe now. I promise you, from the bottom of my heart, no one will ever hurt you again." I never dreamed that I would have to break that promise.

II

As a protection against predators, gibbons in the wild seldom set foot on the ground. Most of their food is found in trees and when they drink, they swing from an overhanging branch and reach into the water with a long-fingered hand to scoop the liquid into their mouths. It seems strange, when gibbons can walk entirely upright, that they don't run on the ground, but predators are many. Safety is in the trees.

At home, once we had run the gamut of curious servants, dogs, cat and daughters, we took Penny to Van's and my bedroom. I undid her leash and let her go.

After a minute or two, and without looking at any of us, she stood up and tiptoed into our dressing room. For the first time, we could see that she stood about ten inches high and had been injured in one leg. Instead of walking straight forward, she danced sideways in a graceful little step-slide, one arm raised above her head as if she were performing a Polish Mazurka.

My closet door, in the dressing room, was open and Penny climbed my dresses until she reached the shelf. Higher than my head, she seemed to feel safe and made her way to the farthest and darkest corner. In a few minutes the lids closed over her shoe-button eyes and obliterated all the terrors of her day.

The bathroom joined the dressing room and we set a dish with cucumbers, bananas and rice on the blue tiled floor in hopes that when it was quiet she would come out to eat. There was plenty of water in the bathroom. At odd hours during the day the water went off in Bangkok. In the event that you were lathered with soap when that happened, some houses had huge jars which were filled with water and a tin bowl to scoop it out with. Our bathrooms had a built-in, tile "klong-jar" and we were given a silver bowl as a scoop. The jar was always full.

We straggled down the stairs filled with disappointment. We had expected a cuddly, affectionate gibbon like the ones we had come in contact with at other people's houses. Instead we had poor, bedraggled Penny who saw us as the enemy and whose only thought was to get as far from us as possible.

We had brought other animals from terror to trust, and we knew that time and loving care would erase whatever memories Penny had of the nightmares she had lived through, but we were eager to begin her rehabilitation and she had successfully blocked us out. We had to be patient and wait.

Day followed day and by the end of the first week I could pick Penny up without being bitten, although she still hung from the end of my arm. She seemed to be terrified of my body.

The second week she made friends with our dogs, Pruno and Yorick. Tucky, the cat, because he was nearer her size, she saw as a friend and fellow victim. She loved him and would throw herself on his sleeping body prepared to cuddle. Long-suffering Tucky never stayed around to cuddle. He usually got up and stalked haughtily away.

Penny always seemed aware of where I was. If I stirred,

she ran behind a chair or under a table, but after a few days, if I moved to another room, she followed at a distance.

The second week, too, she began to accept food during the day even though she went behind a chair to eat it.

Since there is no twilight in Bangkok—it's light and then it's dark—wild gibbons go to bed at around five o'clock so they won't be caught in the open when predators start their search for the evening meal. Penny established her routine and promptly at five every day, she struggled up the stairs and into her closet. By then we had made a bed for her in her chosen end of the closet, and she had a soft pillow and doll blankets that she didn't need. The dressing room was not air conditioned nor was the bathroom.

Early morning before we were up, apparently, was a busy time for Penny. We had removed or locked up anything that could have been dangerous for her, but she found things that she must have considered wondrous. Her choices were generally messy. The tooth paste from tubes wasn't hard to remove from the tiled floor and walls, but Penny had to be washed too and that was a debacle.

The talcum powder which she sprinkled on, or, in every ledge or crevice was harder to clean up since we had no vacuum cleaner, but we viewed it as a tragedy when Penny scattered the Dutch Cleanser, because this was hard to find in Bangkok.

Luckily, we caught sight of Van's uniform insignia in the commode and didn't flush them away. We'd learned to look before we sat when Lee's doll shoes were observed just as they had disappeared down the drain.

Since gibbons aren't meat eaters, their droppings aren't smelly or frequent. Penny picked one spot—she hung from the wash basin—and only that place had to be disinfected. Since the bathroom floor was washed daily anyway, there was no extra work for the servants.

In the third week we had a real break-through. Penny accepted a banana from my hand, and ate it sitting at my feet. I felt then that I could take her out onto the terrace and let her climb in the bougainvillea vines.

She climbed and swung and chased bugs and tiny lizards called chinchooks in the dappled shade until a car backfired on Pattipat Road.

Like a traveling light beam she was across the terrace and against my heart, her furry, baby arms around my neck. My hand automatically came up as it does for human babies and I pressed her to me.

Neither of us moved for a long time. I could feel her tiny heart thumping in her breast and I'm sure she could feel mine. In those moments, as her heartbeat slowed, I became her mother and she never again, when she was awake, willingly let me out of her sight.

We sat, she and I, until finally she moved her head and, with a soft cooing noise, looked up at me. In her eyes I saw, at last, trust and love. My heart sang!

III

It had all begun in Vienna, Virginia, on what turned out to be the most malodorous and imperfect of days.

We had never thought it would be an ordinary day. Maybe there are people who move a family of five and its pets from a comfortable house in Virginia across the whole of the United States, the Pacific Ocean and the South China Sea to Bangkok, Thailand, on a regular basis, but we'd never done it before. In the previous four months we'd dealt with lengthy lists of unlikely adversities, but nothing prepared us for the unlikely adversities of this day of departure.

Our goal for the day was to reach Long Beach Island, New Jersey, where we would spend two nights with my parents before heading west across the continent.

That August morning in 1956, the sun came up determined to do its damnedest and by 8 A.M. it was ninety-eight degrees, humidity to match.

Van, my husband, perspiration running down his face, jammed the last suitcase into the small, two-wheeled trailer, tied the tarp in place and announced, "We're ready to go. Round up the troops."

As a colonel in the United States Army, where the Military said he would go, my husband went. We—one wife, three kids, two dogs and a cat, all lumped together as "the troops"—went with him. This time, God-willing, we were going to Bangkok.

The troops, all except Tucky, the white cat, were scattered. The large cat had already stretched himself comfortably on the kids' pillows in the back of the station wagon. Convinced he was one of the dogs and, in his own mind, the most important one, he was determined not to be left behind. He also wanted to avail himself of the softest and best spot, so had gotten in the car early to make his selection. Nothing seemed likely to dislodge him, but I clipped the leash to his harness and attached him, temporarily, to a basket that held a large jar of water for the animals, and their feeding dishes. The cat hardly raised his head, but a low rumble started somewhere in his insides and I knew he was content.

Craig, our fourteen-year-old son, came around the corner of the house carrying a large sack of Purina Dog Chow, the only food that didn't upset our Great Dane's inner workings.

"Did you mean to leave this on the back porch?" he asked, already aware of the answer. Van blinked his eyes slowly, drew a deep breath and started untying the ropes of the tarp. When the contents of the trailer were exposed, I couldn't see where there was room for the dog food. It was a very small trailer. I started tugging at one of the suitcases to make room. My husband shoved me aside.

"Now let's lay down the rules before we start out," he said. "Things aren't going to be easy with this motley crew, so let's simplify it. There can be only one commanding officer and I'm it. I'll handle the logistics and you handle the troops. Above all, keep your sense of humor."

I wanted to protest that he had the easier job, but

thought better of it thinking that, as soon as we were under way and cooler, the rules would change. I was wrong.

"I'll get Lee," I said to Craig who had sat down on the front steps. "Have you seen Christy?"

"I think she's in the barn." Craig referred to the beautiful old structure that had been built when the house was built in 1830. I groaned. Nothing in the barn was clean and our youngest, eight-year-old Christy, had a definite predilection for dirt.

"Get her, please." I looked at Craig. His blue eyes stared back.

"You know she'll be dirty."

"Then clean her up and please don't leave the bathroom in a mess. The renters are moving in this afternoon." Craig started off as Van unloaded the trailer.

I walked across the front yard where I could see Lee carrying her doll, Tommy, wandering among the trees and bushes. Tears rolled down her cheeks and splashed on Tommy's bald head. As I came near I heard her quavering voice bidding good-bye to the immense trees planted when the house was young.

"Good-bye, my beloved maple (or linden or horse chestnut). Good-bye, dear oak tree." She caressed the oak's bark lovingly. She'd spent many imaginative hours with her toy horses and dolls among the great tree's roots. "Wait for me. I'll be home in two years."

Pruno, our gentle and gentlemanly Great Dane, and Yorick, a frowzy-furred medium sized dog whose black body held a golden heart, followed our middle child on her pilgrimage. They, too, loved the familiar trees and shrubs and though they had no idea what was in store for them, they followed Lee, sniffing and watering as if in farewell.

Lee and Pruno.

Christy and Pruno.

Van delivers a lecture.

"Come on, Lee-Lee (the pet name for our ten-year-old). We're ready to go."

Lee's expression was stricken and the tears seemed to magnify her already large hazel eyes as she turned to come toward me. Poor Tommy, in her arms, was soaked.

I understood my tender child's sadness. As the daughter of an army officer, I too had left many places and friends that were important to me. I had sometimes felt that my heart would break with the partings, but now I knew that it had only grown larger to accommodate so many wonderful memories and to embrace friends all over the world. My children would learn the joy of adventure and the tremendous wealth of remembrance, but today, for Lee, that knowledge hadn't yet come.

Yorick and Pruno, as different in size as Mutt and Jeff, came galloping across the lawn. Lee, brown head drooping, moved slowly and reluctantly. I put my arm around her shaking shoulders and guided her toward the station wagon. Anything I might have said to her had already been said. Now she just had to know that we understood her sorrow and that she had to accept her fate.

Lee, by choice, had elected to share the back deck with Pruno, Yorick and Tucky. It was a large station wagon, but in spite of the fact that he tried to take up as little space as possible, a Great Dane is a large dog and Pruno wasn't entirely successful. Still, when they were all in and arranged, no one complained and I turned my attention to Craig emerging from the front door towing a protesting apparition.

Christy's brown hair was glued to her head with perspiration. Her hands and face were semi-clean. Craig had done his best with a difficult subject, but Christy's legs and clothes looked as if she'd been wandering the earth from time immemorial and had never encountered any water.

"Van," I said as my husband finished fastening the last tie-down on the trailer, "I don't know whether this comes under troops or logistics, but we have to get Christy some clean clothes. We are going to start out clean."

Van spun around and looked at his daughter. After a long pause, he turned and, with great deliberation, began to unfasten the ropes again.

"I suggest," he said, "that since she comes under 'troops,' but her clothes, because they are in the trailer, might come under 'logistics' and . . . since I will untie these ropes only this once more . . . that we dress her in a bathing suit and hose her off at every rest stop."

The subject under discussion wilted, and though I felt Van might have the perfect solution, I vetoed his proposal. Grumbling, Van found and handed out the requested clean clothes. I stripped Christy to her underwear, miraculously clean, handed her father the dirty shorts, shirt and socks, and escorted her into the house and to the laundry tub in the utility room.

She was a very small eight-year-old, having been born three months prematurely, and was able to get in the laundry tub for a bath, leaving the pristine bathrooms untouched.

Van was standing beside the trailer when we returned. In his hand, the ropes jiggled like a fistful of writhing worms. "Are you going to need anything else?" he asked grimly. "I told you, I won't tie and untie this thing again."

"Where's your sense of humor?" I knew it had gone where mine had, but I asked just to remind him of our earlier agreement.

He looked at me and smiled faintly. "You're right, he said. "Let's keep it at least until we're out of the driveway." He shook his fistful of ropes. "Put everybody in the car so nothing else can happen and then, and only then, will I secure the trailer."

I closed the tail-gate on Lee, Pruno, Yorick and Tucky and went around the side where Christy was climbing in over piles of broken plastic toys, wheel-less cars, rags, boxes of broken crayons and other debris that had been, until an hour ago, in the trash cans out by the barn.

Craig was standing beside the car waiting to follow his sister. "I can't ride with all that junk under my feet, Christy. Get it out of there." Craig began to scoop up an arm load.

"It is not junk, and don't touch it!" Christy had a high piping voice and in her frenzy I know she pierced E above high C. "They're my treasures!"

"Christy." When I'm irritated my voice gets as low as my daughter's was high. I might even have sounded like a man as I continued, "I am going to put this trash back into the can designed for just this sort of thing, and I don't want to hear a peep out of you. Ah...," I said as she opened her mouth... "not one peep." The mouth remained open but, fortunately, nothing came out. "Craig, help me pick up anything that's broken. We're making a clean start." I underlined the word "clean" and glared at Christy. She glared back.

Craig snorted and kicked his sister's favorite doll (aptly named "Poor Pitiful Pearl") as he climbed in to gather an arm load. The usually callous mother snatched her bedraggled child and clutched it to her bony breast while her brother and I cleaned the floor under her feet. We made our trip to the trash cans and returned to the station wagon. Van was standing by the driver's door guarding the inmates. My son and I took our assigned seats.

"Everybody ready?" Van's steady voice had a calming influence. "Yes," we all answered.

"I'll lock the house and we're off." Van ran up the steps, turned the key in the lock and came back to slide onto the driver's seat.

"Ready?" he asked again as he slammed his door. There were three "yes"es and one, "I have to go to the bathroom."

"You don't have to go to the bathroom, Christy." Van knew that his youngest had the capacity of a septic tank. "You had your chance. Now hold it." He put the car in gear and we rolled out the gates, the trailer bumping along behind. I think we were all too tense and hot to even say good-bye to our dear house.

Aside from the challenge of spending two years in a country whose culture was so different from our own, we had the practical problems of taking clothing for three growing youngsters, clothing for a reportedly endless social life for Van and me, a refrigerator, a wringer washing machine, window air-conditioners and all the other things to make life bearable in a country that was *always* hot, steamy and sticky. In addition we were going to cross the Pacific on a luxury cruise ship, the U.S.S. *President Cleveland*. Everything that wasn't necessary on the drive across the country, and the cruise, had been shipped ahead to Bangkok. All the clothing for the five of us for the month-long luxury cruise, had been packed in trunks and sent to Fort Mason, California, to be held there for our arrival.

Our plans had been made and carried out as carefully as possible and as we turned onto Walnut Lane I leaned back in my seat, ready to enjoy a leisurely drive across the USA and an elegant cruise to the Orient. My job was "well done!" Nothing could go wrong now!

"Are you ready for the air-conditioner?" I asked when we reached Maple Avenue, the main street of Vienna. In 1956, cars were not generally air-conditioned and the station wagon was no exception. I had, however, found a wonderful invention at one of the automobile dealers'. It had been advertised and touted on television and in the newspapers, and a friend who had bought one assured me that this was just what we needed to cross the United States in comfort in the heat of summer. It was installed, the kids and I had tried it out and it worked.

It was a tubular steel cylinder about two feet long and six inches in diameter, stuffed with excelsior. It fit onto the top of my partially rolled up window. When the cylinder was filled with water, the excelsior became saturated and, supposedly, when I pulled a string, louvers opened and the air from outside rushing through the wet straw cooled the entire station wagon.

Eagerly I instructed, "Roll up your windows."

The windows rose. I pulled the string. I turned to smile at the expectant faces in the rear.

Suddenly—with the roar of a speeding motorcycle—two gallons of sun-heated water streamed past me and inundated the eager assemblage behind me. Expressions changed and though the bodies seemed immobile, the vocal chords functioned loudly and well.

Tucky reverted to a water-hating cat. With flattened ears and soaking fur, he expanded his tail, extended his claws and launched himself forward onto my red linen lap. He came by way of Craig's head and my shoulder. We were immediately soaked, tufted in white fur, and in pain. Van, who couldn't tell what had happened, swerved to the curbing and stopped the car. He looked at me and then to the rear.

"Is that the way the thing is supposed to work?" he asked wonderingly.

"Of course not," I answered shortly. "I could have saved a lot of money by throwing a large pitcher of water on us all."

"Maybe we should have brought a pitcher, Mom. I do feel cooler." Craig mopped at his formerly well pressed shorts and rubbed his face on his driest sleeve. "Christy, stop yelling. Water won't hurt you." He put his hands on his sister's bouncing body and pinned her to the leather seat.

"Ooooo! My clothes are all wet!" Our youngest had a terrible aversion to water, but she was more outraged than uncomfortable.

The occupants of the back deck had caught the major portion, but in spite of the dogs, who were shaking their excess water onto her, Lee hadn't made a sound.

Her dripping arms were held out from her body, her

head was down and her disbelieving eyes were fixed on her soaking shirt and shorts. At last she sighed, and, with her doll blanket mopped her own face and her doll's. Then she looked up and said with resignation, "Well, that's that!"

Van stepped quietly from the car, came around and, without a word, unbolted and removed the malfunctioning cylinder from my window. He laid it carefully on the curb, reappeared on his own side of the wagon and, still in silence, slid in, slammed the door and we drove away.

We had agreed that, if we could keep our sense of humor, the trip could be fun. From his shaking shoulders I could see that Van had found the funny side.

We had to make, my husband said, one quick stop at the Pentagon because a civilian employee, Mr. Wilson by name, had not in four months gotten around to signing the travel vouchers or giving us our steamship tickets. Prior to our trip to Bangkok, all military personnel who refused to fly to a foreign destination had been sent by freighter. Christy's health was not always robust, and freighters carried no doctors. We didn't dare go by freighter. Because of Pruno, Yorick and Tucky, and because I was terrified of flying, we didn't want to fly.

The military ordered a staff study of the situation and found that should Christy become ill and the freighter have to turn around or put into a foreign port in order to find a doctor, the Government of the United States would be liable for the cost of the detour. We were to be sent aboard the U.S.S. *President Cleveland* which was all we'd asked in the first place.

On this, our day of departure, Mr. Wilson had promised to have everything ready by eleven A.M.

We drove into the parking lot at the five sided office building at 10:45. Van parked us near the only available tree, a three-year-old sapling with just enough leaves to show that it was going to be a maple.

"I'll be right back," Van promised as he went off toward the air-conditioned building.

At 11:30, after the four of us had grown tired of watching the heat rise from the pavement and were out of the wagon and clustered around the sapling. Van came stomping across the parking lot. His lips had grown thinner and his jaw twitched.

"Mr. Wilson," he said grimly, "has gone to an early lunch, but is expected back shortly. He was gone when I got up there at 10:55. I'm sorry!"

As his father strode off again Craig said, "I could almost feel sorry for Mr. Wilson when he and Dad finally meet, if I weren't so hot. As it is, I hope he gets what's coming to him."

At noon, Van was back with a box of hot dogs cooler than we were and cokes that were as hot.

By 12:40 the kids and I had just put the finishing touches to a story about Mr. Wilson's enforced retirement when Van came toward us waving a manila envelope.

"Hooray!" we cheered, and even the hot, panting dogs gathered the strength to stand up and wag their tails. Tucky, on harness and leash, lay under the tree, totally noncommittal.

"You don't get to celebrate yet," Van warned. "I have the vouchers to get the tickets but not the tickets themselves. Mr. Wilson forgot."

This time I mentally fired Mr. Wilson without giving him the option of retirement, and hoped Van had attended to the same end in the hours he had waited in the Pentagon. Mr. Wilson must have had a boss and I felt sure that the furious colonel standing now in front of his bedraggled family must have consulted with him. I thought, however, that I'd ask later when things were more calm.

"What now?" I asked as we rose from the hot curbing and got into the scorching car. The leather seats burned through my dress and as Craig and Christy screamed, I sympathized with them for the pain on their poor bare legs. Lee, on her pillow in the rear, was all right, and she quickly tossed the other two cushions to her siblings. "Ah!" they sighed in unison and settled back.

"You asked, 'what now' before the furor in the back began." Van's lips had grown even thinner and his jaw was a cluster of muscular knots. "Now we go to 15th and K Streets in Washington." He put the wagon in gear and we rolled forward. "There, if we can find a parking space for a station wagon and trailer, we'll also find the President Steamship Lines." Van, in spite of being an army intelligence officer, was usually an optimist. I could tell that the day and Mr. Wilson were beginning to alter his outlook on life.

On the 14th Street bridge, I turned to take inventory of the passengers in the rear. We were not the same clean, well dressed family who had gotten into the car some hours earlier. All clothes, mine included, were wet with perspiration, and grimy with dirt that clung to the soggy fabrics. Our hair was plastered to our dripping heads and the kids, after mopping at their faces with filthy hands, had streaks and lines that made them appear to be unwanted and unwashed ragamuffins. Only Van, who'd been in an air-conditioned building, still looked like, to use Christy's phrase, "a human bean."

Even the panting animals, including pure white Tucky, looked damp and grimy although the cat was working on himself.

No one was exactly chatty as we crossed the Potomac River into the District of Columbia.

On K Street, when two cars pulled out in tandem and left room for the station wagon and trailer, Van shouted, "Hallelujah! Our luck is changing. This time I really will be right back." He took his manila envelope and went into another air-conditioned building. I felt unreasonably resentful. Looking as I did, I wouldn't have gotten out of the car if he'd offered me the chance, but the temperature had to be nearly one hundred and I wanted the offer.

"At least it's shady and that's an improvement." Craig must have read my mind. He peeled off his shoes and socks and put his feet up on the back of the front seat. As I turned my head to look at him, his big toe nearly put my eye out. I glared at him.

"Sorry," he said as he put the feet on the floor.

Christy, already barefoot, turned upside down and, resting on her shoulders with her back against the seat back and her legs in the air piped, "Why isn't he back yet? He's been gone a long time and he said he'd be right back. He's a liar. Daddy is a liar."

"Christa Lambert Vanderhoef, don't you ever say such a thing again! Don't ever call anyone a liar, and sit up on the seat. You don't need to make us look worse than we already do." I was irritable.

"Well, he is," came an almost inaudible murmur from behind me. I was too hot to deal with the remark.

"How come those people out there look so nice and clean?" Lee was watching passersby on the sidewalk. "They look cool, too," she added.

"They've probably been sitting in an air-cooled restaurant and are now on their way to their comfortable, air-conditioned offices." I'd never wanted to work in a office before, but now, I felt bitter and discriminated against.

Minute after minute passed. The temperature rose. Soon I began to think Christa had been right. Van was a liar. Each occupant, glued to his own wet leather seat, grew more miserable and silent.

Pruno, Yorick and even Tucky had been panting and had their heads out of the windows. Pruno drew his in and began to pace as far as the limited space would allow. When his mournful howl rent the air, we all jumped.

Fortunately, Lee jumped the farthest because as she leaped into the little aisle next to Christy, poor dignified Pruno went off like a siren from one end and a bomb from the other. The heat had so upset him that he had diarrhea all over the available space in the back deck.

There was a second of horrified silence, then Craig, Lee, Christy, and Yorick joined Tucky and me in the front seat. It was crowded.

One expects a large dog to have a tremendous capacity and a powerful smell, but Pruno became the champion of all time right then.

The gagging, yelping and hissing emanating from the car as the doors flew open and we huddled in an unclean cluster on the sidewalk, caused passersby to stare in our direction. The stench, as it boiled through the openings, forced people up against the buildings.

They hurried by, hands ever their noses and with eyes searching for the source of the terrible odor so they could avoid it. Since Pruno was, by then, out of sight between the front and back seats, his embarrassed head buried in his paws, they probably thought one of us was the cause.

"Do something, Mother, do something!" The girls brought me out of, what might be termed, the anesthetic.

Forgetting that we had already been viewed by nearly everyone in downtown Washington, I made a ridiculous statement.

"Get in the car," I said, "before anyone sees you. I'll take care of it." They all dutifully climbed into the front seat.

Yorick, without leash, was investigating a tree near the car, but came right away when I whistled. Tucky, unnoticed, had wandered down the sidewalk and was the center of an admiring group of office workers. One of them, an attractive young woman in a navy blue linen dress, picked him up. I hated to draw attention to myself, we were getting enough of that already, but I could see that our cat was about to be kidnapped.

Tucky, considering himself a dog, always came to a whistle, but when I had summoned Yorick, I did it so softly that the cat, who was farther away, hadn't heard me. Now I puckered up and blew and was probably heard in Baltimore.

Tucky wrestled free of the embracing arms and flew toward the car, but not before he buttered the front of the blue linen dress with white fur. Somehow it made me feel better about my own fur-encrusted red dress. Serves her right, I thought. I didn't know what the girl had done to deserve my anger, but I think it must been because she had looked so nice when I was in such a mess. Well, now she looked almost as awful as I did.

Once everyone was corralled, I began to root through pile of bags and boxes that had been crowding my feet on the floor of the car. I had bought a giant, economy-sized box of Kleenex that I thought was large enough to last us all the way to California. Naturally, it was on the bottom of the pile.

Bearing it with me, I made my way to the rear of the

wagon; lifted my tight, straight, wrinkled skirt above my knees and straddled the tongue of the trailer.

When I opened the tail-gate and my nose was in closer contact with the disaster, I thought I was going to add a new dimension to the problem.

"I'm going to be sick," I said aloud.

A chorus in the front sang, "No, don't do that," *a cappella* and with feeling. I stepped back, swallowed and crept forward again.

Using the Kleenex and what was left of the animals' drinking water, I sopped and mopped until the floor looked clean. The odor, however, remained. Van's thermos of coffee went next. There was no improvement. I emptied the shoe box that held the girls' crayons and filled it with soiled tissues. I also filled the box in which Van had brought the hot dogs, two large paper bags and the Kleenex box itself. That collection, for want of a waste basket within running distance, I pushed under the car. Still the putrescence clung.

Desperate, I grabbed a huge plastic bag of bright green, granular bubble bath that a neighbor, bless her, had brought the girls as a farewell gift. Sprinkled thickly on the offending area, it smelled strong and cheap like dime store perfume, but it had more power than Pruno's calamity and to us it smelled wonderful. I blanketed the back of the wagon with it.

"I like that." Christy, who had been leaning over the back of the front seat, smiled. "Dad's gonna like it too." I didn't agree with her. "Going to," I automatically corrected.

"It's better than the diarrhea." Craig got out and changed to the middle seat of the wagon. "It's strong but it doesn't make you sick at your stomach. It'll probably weaken as we go along, won't it?" I said I hoped it would.

"Well, I can't sit back there in the bubble bath. I'll sneeze all the way to New Jersey." Lee had a point. "May I sit up here with you and Dad?"

"Yorick, get in the back." The dog's long lashed, brown eyes stared at me pleadingly out of his sparsely furred, bony skull, but he obeyed. "Lee, give Tucky to Craig and slide over." The cat was handed over the seat back, and my daughter's bare legs squeaked on the damp leather as she wriggled into her allotted space. I plopped onto the seat and shuffled my feet into the clutter on the floor. When they touched bottom, I sighed. That crisis was past.

We were barely settled when Van emerged from the building. His step was jaunty and he waved the envelope at me as he approached the car. Halfway across the broad sidewalk he raised his nose in the air and began to sniff.

"He smells us," Christy said happily. How could he not, I thought. "He thinks we smell delicious," Christy continued. I marveled at her innocence.

Van's first words, "What's that awful smell?" deflated his youngest like an overdone soufflé. At her expression he thought he had found the culprit. "Christy, have you . . . ?"

"For once Christy didn't do it, Dad." Craig was usually fair. "Pruno. . . ."

"Pruno had diarrhea," Lee chimed in. "Mother cleaned it up but. . . ."

"She used up all of our bubble bath." Christy had found a bright spot. "Now we can't take a bath," she added.

"Fortunately, you can and will." I answered. "Accept this fragrance, dear one," I said to my husband as he slid behind the wheel. "The one it is masking was unbearable. Do we have the tickets?"

Van reached across Lee and handed me a bulky envelope. "Put these in my briefcase, please. He looked at me. "Where is my briefcase?"

"It's back here, Dad," Craig volunteered. "Pruno is lying on it."

"Why isn't he in the back?" This brilliant army officer didn't seem to be getting the point. He looked at Lee sitting between us. "Why is everyone out of order?"

I explained our recent disaster once more.

"Oh," he said when I had wearily gone again through the corrective steps I'd taken. "Well, the animals' tickets were the things that took the time. At first, no one would believe that we were crazy enough to take this kind of a luxury cruise with three kids, two dogs and a cat." Amen, I thought. "Actually, the man and I came out onto the sidewalk to view the menagerie, but there was a terrible smell so we only stayed long enough to see you standing behind the car with your skirt half-way to your hips. What in God's name were you doing?"

I couldn't believe that he still didn't know what I had been doing.

"We," I said quietly, "were the source of the terrible smell that sent you back into the building. "I," my voice rose, "was back there cleaning the car and over-riding that smell with the flowery fragrance you now dislike."

At last Van got the picture. He reached across Lee to pat me on the knee. "I'm sorry," he said in apology and sympathy. "I didn't realize." Then he chuckled. "If it's any consolation, the President Lines man and I thought the view of your legs was great."

Surprisingly, the little compliment did help my morale and as we wound our way through the, by then, evening rush-hour traffic, I smiled.

Even Pruno, who still lay under Craig and Christy's feet

hiding his shame, crept up onto the back deck where he sat, rather unhappily, in the fine, cloying crystals.

I handed the ticket envelope to Craig who put it in the released briefcase.

When we got to the Washington-Baltimore Parkway and were up to speed, the wind from the open windows swirled and shifted the powdered bubble bath like the sands of the Sahara.

I watched as it covered the kids' sweaty bodies with green. They watched as it covered ours. It stuck like glue. It dyed our hair, it provided us with three green animals and—worst of all—it blew in our eyes. Being salty, it stung! Van couldn't see through his tears. He wavered to the side of the road and came to an abrupt halt.

"We can't go on like this!" bellowed my usually placid husband as he mopped his streaming eyes. "You have to do something!"

"Why do *I* have to do something?"

"Because you're the mother." He looked at me, his green face determined.

"That's all I've heard all day." I said each word carefully through clenched teeth. " 'Do something, Mother.' It seems to me that this comes under logistics. Besides, I'm not your mother. You do something."

"I don't know what to do." He sounded pitiful. My anger subsided.

In the glove compartment was a spray can of Fuller Brush Insect Repellent. I reached in, grabbed it and handed it to Craig.

"Spray the stuff. Wet it thoroughly. Stick it to the floor," I growled. "I hope this works. If it doesn't, I have nothing further to offer."

An aroma, redolent of an old trash dump, wrapped itself around us as Craig sprayed, but the repellent glued the granules to whatever they lay upon, and our driver was coaxed back into operation.

The remaining five hours of the trip seemed to stretch beyond eternity. We couldn't stop at a restaurant lest the other patrons think that smelly Martians had landed. We ate our few Oreos and Fig Newtons and sucked on Lifesavers in lieu of water.

At last, we drove into my parents' driveway on Long Beach Island, New Jersey. We were hot, tired, hungry, thirsty and, as Craig said, thoroughly "un-gruntled."

As our station wagon rolled into my parents' drive, my father and mother rushed out to greet us but just when I was ready to throw myself into loving and comforting arms my father clapped his hand over his nose and backed away.

"Good God! You stink!" he yelled loudly enough to alert all of Long Beach Island.

Mother looked bewildered, but she stood her ground. I did the only thing possible, I burst into tears.

"Everything stinks!" I yelled at him through my tears. "And this day stinks most of all!"

I got out of the car and threw my soggy purse at nothing in particular. It landed somewhere in the dark yard.

Having been on duty for twenty-one of my thirty seven years, my mother recognized the symptoms and was prepared to "do something."

That precious lady ignored my smell, opened her arms and took me and my responsibilities into them.

In that moment, as my obligations shifted temporarily from my shoulders to hers, the day, that had long since turned to night, and I were all right again.

IV

Late in the afternoon of the third day on the road after leaving my parents' home, we drove up in front of the beautiful home of Van's Uncle Will Nash, near the small town of Wayzata, Minnesota. The house stretched its wings along the shore of Lake Minnetonka, and we thought two days there would be a welcome break for people and animals. We reckoned without Uncle Will's dachshund, Vanderhoef.

Aunt Laura, Van's aunt, had been a Vanderhoef, all four of the Nash children had the middle name of Vanderhoef and there was always a dog of some breed to carry on the name. The current bearer of the standard was the snarling, barking dachshund who attacked the car as soon as it stopped rolling.

Yorick and Pruno, who had been eager to disembark, withdrew their heads from the windows and looked at us as if to say, "We don't think we're welcome here! What now?"

I wondered the same thing, but a soft voice with a Norwegian accent silenced the feisty dog.

"Stop dat noise, Vanderhoef." Miss Nelson, better known as "Uh-Huh," came around the corner to greet us.

Vanderhoef stopped barking, but none of us got out of the car.

Uh-Huh had come to the Nash household many years before when the four children were small. She stayed as their nurse until they were grown and then became the housekeeper and a true and beloved member of the family. Aunt Laura had died before Van and I were married, and tiny Uh-Huh ran the household and its inmates from that time on.

"Dean," she said to my husband, poking her head in the window and using his given name, "Vanderhoef doesn't like udder doks, so ve haf arranged wit a kennel. Edgar iss coming now to take you der." Edgar was Van's older cousin, the eldest of the Nash offspring.

"Jeannie!" Uh-Huh smiled and put her hand through the window to touch me lovingly on the arm. "You and the kits come inside de house. Bring de ket wit. Vanderhoef issn't jealous of kets."

Edgar drove up with his wife, Nancy, and they both got out of their car as the children and I, Craig bearing Tucky, extricated ourselves from the possessions and trash we had accumulated in just a few days.

We loved Nancy and Edgar and after we'd all hugged several times around, Edgar said, "Come on, Dean, we'll take the dogs to the kennel," and off they went.

I wasn't comfortable about Pruno and Yorick being in a strange kennel. They'd been uprooted from their home and I worried that being torn now from the family in a strange place, might turn Pruno back into the quivering, cowering mass he'd been when we found him in Germany. Still, there was no choice.

Nancy and Uh-Huh told us what a loving, caring kennel it was and how careful the staff was with the animals it boarded, and I, at least for the time, felt better.

We had a lovely dinner that night. The other cousins,

Bill, Fred, and Marianna, came with their spouses and their children. The younger children were fed early, and the ones old enough to he civilized joined us at the table in the enormous dining room. It was fun, but the dogs were on my mind and I called the kennel right after dinner. I was told they had eaten well and seemed content. I relaxed.

After the kids were all taken home to bed, we met at Fred's house and were introduced to his raccoon who, because it was night, was wide awake and ready to entertain us. We were entranced and I made up my mind that someday I'd have a raccoon.

Van called the kennel after breakfast the next morning. The dogs were fine, he was told, but we had just finished lunch when the phone rang. The caller asked to speak to Colonel Vanderhoef. Since we knew no one in Minneapolis, we knew the call had to be from the kennel, my parents or the Pentagon. No one else knew where we were.

When Van came back to the dining room, his face was pale. "Come on," he said, "the dogs have escaped and are loose in Wayzata. Sorry, Uncle Will, we'll be back when we find them."

Lee and Christy, after a busy morning of swimming with their cousins, had eaten early and were napping, but Craig jumped up to join us. As we hurried from the dining room, I heard Uncle Will say to Vanderhoef who was lying in the dining room doorway, "See what you caused, you smelly, old, bad tempered dog."

In the car heading toward the established rendezvous, I asked Van how the dogs had gotten out.

"Pruno stood up and opened the latch with his nose and out they went. There was no lock on the gate. Fortunately the owners saw them go and are in pursuit."

It took us ten minutes to reach the command post, a

lovely residential area of upper middle-class homes and huge shady trees. There we found the animal warden, his truck, his assistants and an ever increasing crowd of dog-lovers. By now the kennel owners and the entire kennel staff were in on the chase. The dog warden, two veterinarians and assorted members of the Great Dane Club of America had been rallied to arms. We also found that the dogs had been on the loose for some time and it seemed that everyone in the state had been called before we were alerted.

The police, contacted by Uncle Will, showed up just as we did. The quiet village of Wayzata had never seen such excitement! Like the Keystone Kops, the army ran through the town trampling flower beds, peering under bushes, into windows, doors, sheds and garages, scarring lawns and wreaking havoc.

"I'm gonna sue the city," an irate man bellowed as he trotted beside a large, black Great Dane attached to the dog warden by a stick with a noose on the end. The noose was around the dog's neck.

"He's right," I said to the warden as I joined them. "Our Great Dane is brindle; this one is black. This isn't the one you want."

"Who are you?" the warden asked haughtily, still pressing forward toward a large truck with cages in the back. The truck was old and looked dead tired.

"I'm the owner of the dog you're hunting, and believe me, this isn't the dog."

"The report said the dog was black." We were nearing the truck by now.

"There are two dogs." I explained. "The smaller dog is black. The Great Dane is brindle."

"He took Caesar right out of his yard . . . right out of his own yard!" The dog's owner couldn't have been more than five-foot-six, had a belly like the beer barrel it probably came from, and, I noticed, was wearing bedroom slippers.

We stopped at the truck and the warden prepared to hoist the Dane into one of the cages. The dog's front legs rested on the tail gate, the tail wagged, but the hind legs only danced on the pavement. Caesar turned his head and washed the warden's face.

I wanted to get on with hunting Yorick and Pruno, but I couldn't let this injustice continue.

"Look," I said, "we're visiting Mr. Willis Nash. Our name is Vanderhoef. Our two dogs escaped from the Deerwood Kennels and I really would like to get on with hunting them. Believe me, this is not the dog you want."

Uncle Will's name, and the facts, finally got the warden's attention. After a minute's thought he said, "Well, Okay. Here, Mister, here's your dog, but you'd better lock him up or somebody else is liable to take him again."

The black Dane's owner became less apoplectic as he took Caesars's collar. Caesar joyously moved his fore feet from the truck to the little man's shoulders and in fits and starts they moved across the pavement. The man was slipping and staggering in his bed room slippers and the dog, walking upright, was at least a foot taller than his owner.

"I hope you find your dogs," the man managed to gasp as I hurried past. "I know how I'd feel if it was mine and only these idiots were out hunting him." He gave a short laugh. "We sure get to love 'em, don't we? Well, good luck!"

As I hurried away I thought about my love for the two missing members of the family. I remembered Pruno when we'd first acquired him in Germany. Mistreated and abandoned, he'd been a huge, quivering mass of bone draped with loose skin. He had loved us at once, but for more than a year, a strange man or a loud noise would send him shivering into the farthest, darkest corner. Now, with love, he was a beautiful, happy dog and we loved him with all our hearts.

Yorick had been one of a litter born under the coffee table in our TV room in Vienna, to mother's poodle, Folly. He was the most endearing, if the most wispy looking, puppy of the litter and he too dwelt in our hearts.

I was determined that even if Van had to retire from the army and we never got to Bangkok, we would stay in Wayzata until both dogs were found.

Van and Craig appeared on either side of me and swept me along with the ever increasing crowd.

"Look, up there, Jeannie," my husband said suddenly. "Isn't that the boys?"

Far in the distance I saw two specks cross the road. They were of uneven size and each had four legs. "It does look like them, but I'm afraid to hope."

"Whistle, Mother," Craig said. "Maybe they'll hear you."

I can summon the loudest, shrillest whistle in the family when the occasion demands it, so I put my lips together and blew. The specks hesitated. My heart leapt and I blew again. They stopped. I whistled once more. They turned and raced down the road toward us. No grasping hands or prone bodies deterred them. We began to run and the meeting came somewhere in the sea of floundering, applauding humanity. When the dogs reached us, tongues lolling, sides heaving and tails waving, they leaped upon us in unalloyed joy. Pruno stood half a head taller than I when he put his paws on my shoulders and slathered my face. Yorick sprang into Van's arms and there wasn't a dry eye on the block when the reunion was over.

Van made a touching speech of thanks to all the city employees and volunteers; the dogs jumped gratefully into the station wagon, and we drove away. Our prayer of gratitude came from the very bottom of our hearts.

𝒱

"What country are we in now?" Christy's piping voice wafted forward from the back deck of the station wagon. All three offspring were stretched out sideways on the floor, their heads on their pillows feet prominently displayed in the long side windows.

Yorick lay between Craig and Christy, his chin on our son's chest. Pruno had been promoted, to his way of thinking, to the middle seat, and as I turned to look at Christy, the dog's feet jerked, his ears twitched and he gave a muffled bark. He was sound asleep.

Lee had made a bed for her doll on an extra pillow, but Tucky found that a waste of bedding. The seventeen-pound cat lay upon the doll's unresisting cloth body and poor Tommy, the doll, was smothered by a blanket of white fur stuffed with sinew, bone and muscle. Tommy's mother and Uncle Craig appeared to be asleep.

We'd been underway from Virginia toward San Francisco for two weeks and had settled into a, more or less, routine when we were in the car. Part of the routine (like the inevitable "Are we almost there?" from our youngest) was the unavoidable and just asked, "What country are we in now?"

Author and Tucky

Van and I suspected that Christy already understood the answer. She also had enough intelligence to see that, like Chinese water torture, she would he able to slowly drive her parents to the brink of insanity and, quite possibly, assist them over the precipice.

"We're still in the United States," I said to our youngest with, I'm sure, an edge to my voice. The muscles in Van's jaw, when I looked at him, were jumping as if he had a tic. "I don't think I can go through this again," I muttered.

"I'll give it a try." My husband took a deep breath. "Now listen carefully, Christy," he said, "and don't forget what I tell you. We will be in the United States of America until we get aboard the ship. Isn't that what Mother told you yesterday?"

"And the day before and the day before that, ad nauseam," I mumbled.

"Yes, but. . . ."

"In all the time we've been traveling have we left the United States?"

"Yes."

I could see Van's lips move as he counted to ten.

"What do you mean, 'yes.' When have we left this country?" His voice, when he spoke was very, very quiet.

"Grand Forks isn't in this country," Christy said triumphantly.

Van's voice began to rise and to develop a keen edge. His daughter was impugning his home town in North Dakota.

"What do you mean, Grand Forks isn't in this country? Of course it is. What makes you think it isn't?"

"When we got to Auntie Lois and Uncle Bud's, you said we were in God's country. So now... are we in God's country or our own country? What country are we in?"

"Mesopotamia!" Craig answered irritably and sat up.

"Where?" His sister was interested.

"Mes-o-pot-a-mia!" Craig pronounced slowly.

There was silence as Christy appeared to be thinking. I doubted that she really was.

After a minute or two she lay down, and said, "When we get to Yellowstone Park, I'm going to climb Old Faithful."

Our daughter had a thought all right, but it was a totally unexpected one. Since we had spent the day before on the difference between states, countries, counties, towns, mountains and geysers, I hoped we weren't going to get into that nerve-knotting discussion again. Apparently everyone else was holding his breath at the prospect, because we all exhaled at once and the car seemed to shiver.

When we reached the entrance to Yellowstone Park,

the park ranger was friendly, polite and helpful. He filled the front seat and my lap with leaflets and brochures, told us that since it was the second of September, all hotels in the park were closed for the winter (having closed on the first), but he assured us that we would be comfortable in one of the log-cabin campgrounds. Van was thanking him and preparing to drive away when Christy leaned over the back of his seat.

Her short legs waved in the air behind her, and she stuck her head out of the window almost into the ranger's face, announcing in an excited voice, "We're in a mess of pots, and I'm going to climb Old Faithful as soon as we get there."

I don't know why I felt obliged to explain. "She thought we were still in God's country instead of the United States," I said foolishly, "so our son told her we were in Mesopotamia." That wasn't at all what I meant to say and it certainly was no clarification.

The young man's smile shrank. His eyes, a clear blue only moments before, became opaque. He stepped back from the car and from Christy's upturned face, and said, more to the animals than to us, "Be sure you have your pets on leashes and under control at all times. We have bear in the park." He looked toward the dogs and cat clumped together in the rear, and I was sure he was speaking to them rather than to us.

Craig and Van began to laugh at the same time. Lee and I joined in as we rolled away, our guffawing echoing through the trees. Only Christy was bewildered. When I looked back, I saw the ranger still staring uncertainly after our station wagon.

"That certainly was an edifying dissertation," Van

chortled when he could get his breath. "It proved, without doubt, what that young man had only suspected."

"Sort of like Mother's beautiful Palominos." Craig doubled with laughter again. He referred to an inane statement of mine when, once, someplace in the middle of Illinois, I had seen what I thought were graceful cream colored horses in a lush green field.

"Look at the beautiful Palominos, kids," I'd said with enthusiasm. I wasn't wearing my glasses and I'm quite nearsighted. There was a silence while everyone looked.

The silence was pronounced enough to cause me to snatch the glasses from my lap and put them on. What I saw were four, dirty, white mules standing near a stock tank, in a field that at best was only patched with green.

The laughter, mine included, was loud, and "Look at the beautiful Palominos" became a favored family observation.

At last the laughter diminished, and as we wound through the beautiful park, a family of bear beside the road, a mother and two cubs, entranced us all and I hoped Christy had forgotten Old Faithful. Her occasional remark on the subject, however, kept us aware that in spite of any arguments we might have offered, she was bound and determined to climb the geyser. The wild animals seemed secondary. I dreaded her disillusionment.

The cabins, when we found them late in the afternoon, were primitive. They were so small we required two. Fortunately, they were close enough together that we could put the kids in one and we could take the other and still hear almost every breath they drew. Unfortunately we could also hear the breathing of the man on the other side of us who, once he had coughed and cleared and snorted and snuffled his way to bed, went immediately to sleep and to snore.

The floors of the cabins were oiled linoleum. There was room in each cabin for a pot-bellied stove, a small table, two straight chairs and a double bed. We picked up a cot for Craig which crowded the kids' room somewhat, but if he had shared the bed with the girls, it would have been like sleeping with a pony in a dresser drawer and besides, it wasn't suitable.

It was cold and when bedtime came, we made no attempt to open either of the two small windows that flanked the front doors. It was just as well, as Pruno spent the night with his front feet on the window sill growling at the bears, and we could hear Yorick's echo from the kids' cabin. We knew the dogs weren't disturbing our next-door neighbor because his racket never ceased, so we made no attempt to quiet them.

Tucky huddled under the covers with me while Van and Craig spent the night adding sticks of wood to the stoves.

The communal bathrooms and showers were fifty yards away and no one, not even the chain saw next door, braved the wild animals and the cold during the night.

Dawn came, the cabins finally warmed and we all went to sleep only to be wakened in a few minutes by clamor and hubbub directly under our windows.

"Would you look at that!" A female voice, as high and piercing as an aging diva's, brought us bolt upright. The snoring next door stopped abruptly.

"Boy! Are they going to be unhappy when they wake up and see this." A younger male voice chimed in.

"Wouldn't you think they'd have had better sense than to leave this stuff out here?" Another male voice.

"I hope they clean up the mess. The bears sure won't do it." A convention was obviously convening.

Van and I wrapped ourselves in a blanket and, like Siamese twins, hobbled to the window. Pruno was already there, ears alert, tail wagging.

The window wasn't large, but it was big enough to frame a disaster. Our tiny trailer was surrounded by fellow campers and more were hurrying toward it. A can of dog food lying outside the awed circle gave us a clue.

"A bear has eaten the dog food," Van said. "That's all it is."

"I hope that's all it is," I mumbled. The crowd shifted and I saw a suitcase on the periphery of the circle. "I hope they didn't damage our clothes."

Neither one of us knew then how hard it was going to be to replace that dog kibble, the only kind that didn't upset Pruno's delicate inner workings.

By the time Van and I got our clothes on and had elbowed our way through the milling mob, the kids and Yorick, firmly attached of course, were already on site. Tucky peered through the window of the cabin, sitting, we assumed, on the small table under the window.

The trailer had suffered a definite catastrophe, and it must have been noisy. Obviously we had slept more than we thought. The tarp was more like canvas fringe. Great claws had combed easily through it, and it hung around the edges of the trailer like a threadbare hula skirt. Luggage was scattered on the ground in a wide arc as if the bear had been using each piece as a football. The Purina Chow, naturally, was gone, and the condition of the huge sack that had held it attested to the extreme hunger pangs of its consumer. Cases of canned food, both dog's and cat's, demonstrated the efficiency of bear claws as can openers, and only one or two cans had escaped notice. I wondered how an animal could have known there was food in the cans. I still have no answer.

Some of the crowd stayed to commiserate and to help us restore order, but most of them wandered away in search, I suppose, of breakfast. The luggage was, thank heaven, intact, if a bit scratched and dented. Since our pets didn't eat much during the traveling day, we thought we'd have time, before supper, to find a grocery store and replace their food.

I'm sure that some store in the West carried Purina Dog Chow, but we never found one in any state including Hawaii. For the rest of the trip, Pruno and Yorick ate whatever we could get, and we made sure we stopped often for Pruno's fragile insides.

Once we had tidied the campground we went to breakfast, and the whole time, Christy sang a tuneless song. "When can we go to Old Faithful? I'm going to climb Old Faithful. Let's go to Old Faithful." She had a few variations on the same theme but they did nothing to relieve the monotony not only of her insistence, but of our ever-increasingly irritated replies.

As it always does, the time finally came when we were cleaned, fed, reloaded—and on our way to the famous geyser. Christy bounded all over the interior of the station wagon in anticipation, and the rest of us agreed that we hoped she could attain her goal. We had done everything we could to explain the impossibility of the feat, but our youngest had heard none of it. Now, she had almost convinced us that Old Faithful could be climbed, and we teetered between the laws of the Universe as we knew them, and the Bible which taught that with faith, water could be walked upon. Christy certainly had the faith.

"I want to see her do it." Craig voiced the thoughts of all of us, and was answered by a chorus of: "I do too." The subject of our thoughts was so deep in her own excitement she didn't even hear us.

We parked and got out of the car to walk to the circle around the site of the geyser. The benches in the circle already had a few hardy souls shivering in the morning mist. Christy skipped ahead of us and went right on by the people and benches. As she started up another path, Van ran after her and towed her, writhing and protesting, back to the skeptical family.

"Let go," she yelled. "I have to find it."

"It will be right here in ten minutes," Van said. "You'll just have to sit down and wait for it."

"It can't be here. There's nothing here. How can it be here?"

"It will come out of the ground, Christy." Sensitive Lee already felt the disappointment that was coming to her younger sister. "See where the ground is wet. That's where Old Faithful will come up."

Christy stared at the wet spot in the middle of the large circle. I wondered what sort of vision was in her mind. What did she imagine was going to come up out of the ground? I suspected it had to be either like a flag pole or a hill, but since we had been given no clue and because she wouldn't or couldn't describe what she visualized, we simply had to wait.

Old Faithful was right on time. Christy watched the steaming water spout into the air. She looked from bottom to top and back down again. We were all watching her instead of the geyser.

"What's that?" she asked.

"That's Old Faithful," I said. "It does this every hour."

"I can't climb th-a-a-at thing. It's water and it's hot."

"We know," Craig said. "We tried to tell you."

Christy took her eyes from the spouting water and gave him a withering look. I looked for signs of disappointment and saw none. I saw anger and betrayal and when those disappeared, she shrugged her shoulders, sat down on the bench and said matter-of-factly, "Well, that was certainly a bust. What are we going to look at now?" Her tone challenged us to provide something that was better than Old Faithful.

"How about getting in the car and let's look for some wild animals!" Van sounded enthusiastic and I wondered what his reaction was to Christy's unexpected acceptance. I felt grateful, but wondered if there was a lingering something hidden inside the child that we might never be privileged to help her with. At the same time, I felt disappointment that she hadn't been able to make her climb.

As she skipped ahead of us to the parking area, Van said, "I think she's been putting us on all this time, and we've reacted exactly as she expected us to." From the bottom of my heart, I hoped it was so.

$\mathcal{V}I$

The Pacific ocean was blue, the leaves and grass were a lush green; yellow, red and purple flowers pushed up to soften and blur the earth's edges but my husband's face outside the open car window, was magenta.

"This is no joke!" he announced loudly, his clenched teeth acting like a reverberating sounding board. Instantly the troops in the car became alert and apprehensive. We sat at attention and even the dogs stopped thumping their tails. "Some idiot sent our trunks out yesterday on the *President Garfield*. We have no clothes to wear!"

What he meant, of course, was that the carefully selected clothing for a family of five for a month-long luxury cruise had gone off on the wrong ship.

I stared into Van's blue eyes in his nearly purple face and knew I would have to be very careful. I knew it, but could do nothing about it. "What do you mean, they sent them on the *Garfield*? We're sailing on the *President Cleveland*." My voice was as loud as Van's and twice as shrill.

"I know that, and they certainly know that." The colonel had, I was sure, left the perpetrators in no doubt. "The question is—what do we do now?"

The day was Friday, the ship sailed on Sunday and all that was left over was Saturday. Even if dress in the 1950s had been as casual as it is today, the collection of clothing in our "across the country" suitcases wouldn't have been acceptable on the *President Cleveland*.

In 1956 no one lived in jeans except small boys and farmers. Craig had a pair or two of those. Shorts and slacks for females of any age were for sports only. The girls and I had a few pairs of each. Dressing for dinner aboard ship meant cocktail dresses for women and dinner jackets for men. We had none of those! Special nights required evening dresses. I certainly didn't have any of those in my present luggage. In fact, about the only things we were towing in our little two-wheeled trailer were the forbidden articles of clothing.

Added to the imminent catastrophe, Mr. Wilson, in Washington, along with all of his other omissions, had neglected to get us a Hong Kong visa. We needed one. We had to change ships in Hong Kong and we had a five-day layover there. The station wagon and trailer still had to be sold and we had lunch and dinner engagements with old friends on both Saturday and Sunday. Our time was nearly as limited as our wardrobe.

As soon as Van asked me what we should do now, my mind, which had become anesthetized from shock, began to function as it had been trained to do. Sensibly, I thought.

"It's too late to do anything tonight." I began to gather my purse and belongings together. "Let's get settled and then we can decide what to do."

"The powers-that-be are so contrite they've given us what they call the President's Suite in the Guest House. I hope it's as fancy as it sounds." Van came around the car and got in. "They weren't contrite enough to let us keep the dogs with us; they have to go to the Post kennel, but we can keep Tucky."

I hoped Pruno and Yorick had grown more accustomed to kennels on the trip across the country. They'd been required to stay in several when they were unwelcome as house guests, and had always eaten and slept well. Happy or not, there was no choice.

The "suite," when we reached the Guest House, was lovely. Open windows brought in the gentle, late afternoon sunshine, and the breeze from the bay pushed the filmy curtains aside so we could see beyond the Golden Gate bridge to the meeting of the ocean and the sky. My world straightened out and I knew that everything was going to be just fine.

"We'll be all right!" I said. "I'll call Macy's, reserve a store 'shopper' for tomorrow when they open, and we'll have a go at it."

Macy's, the only store I knew that would have everything we needed, including trunks, promised to be ready for us at nine o'clock on Saturday.

The soft breeze kept up its good work all night, smoothing our "furrowed brows" and soothing our frazzled nerves and knotted muscles.

In the fresh morning we put on our only "appear in civilized society" clothes, ate a strengthening breakfast and presented ourselves at the store of our choice.

We started with the trunks, which were then sent up to the shopping service on the top floor. All day as things were purchased, fitted and sent to the shopping service, a kind lady there packed each item carefully in the two trunks. One trunk was to go to our stateroom; the other, directly across the corridor to Lee and Christy's cabin. Craig was to room with another teen-age boy one deck up. The lady made no errors in her packing.

Van and Craig's clothes came first. Macy's assigned a

tailor to us and as the clothing was selected, he pinned, fitted, sewed, and his results went up to the trunks. Finished, Van and Craig left us for the British Embassy and the Hong Kong visa. They also planned to sell the car and trailer while they were out.

Left alone, Lee, Christy, Frances Short (the shopper), and I raced through the departments. All summer stock had been put away on the first of September, but Frances had it brought back, and the girls and I had a glorious time selecting dresses, shoes, underwear, shorts and bathing suits—all at half price.

At the end of the day, after a personal phone call to the president of our bank, we handed Macy's a check. The clothes were ours!

Macy's final act of kindness was to promise to deliver the trunks to the dock before five o'clock. They kept their promise.

The car was sold (trailer thrown in), the visas were in our passports, and we were free to enjoy a lovely dinner with our old friends. We had clear sailing, we thought.

VII

Embarkation time is always exciting! We had assured that ours would also be confusing by choosing to bring Pruno, Yorick and Tucky with us aboard the ship.

No one denied us our right to the gangway on our arrival at the pier. In fact, as soon as we began our approach, the animal lovers among the passengers and visitors milling about, recognizing the difficulty of two harried parents ushering three children, two dogs and a cat aboard, cleared a path for us.

The non-animal lovers faded into the crowd obviously incensed and horrified. I had the feeling that as soon as they set foot on the ship, they would let the Purser hear about the insult to the dignity and well-being of these insensitive people. Still, I didn't think the indignant ones would demand a refund.

The Purser himself, having heard that a family of five with three appendages was to sail aboard the Cleveland, greeted us by name at the top of the gangway and escorted us, quickly, to the elevator. The attention had less to do with who we were than with getting us out of sight. The kennel, the Purser said as he pushed the elevator button and prepared to depart, was on the sun deck and the steward who would attend our four-footed family members, was waiting there to take over. He then saluted and disappeared.

When the elevator doors opened to admit us, an obese little East Indian man, already on the elevator, paled as friendly Pruno stepped in and advanced toward him. Shaped like the Michelin tire boy, the man tried to wedge himself into the corner at the sight of our gentle dog, but his broad hips and undulating belly forbade it. Pruno's nose almost disappeared into the soft flesh of the man's mid-section before Van could shorten the leash to restrain him.

"Please to remove tiger," the small man bleated in imperfect English. His eyes were fixed on the light in the ceiling as if he thought that by not looking directly at any of us, he would cause us to disappear. Van pushed the elevator button.

Brindle Pruno did have stripes in strategic spots and, in time, we would be indebted to his markings for our very lives, but we weren't accustomed to this kind of mistaken identity and it took a few seconds to understand.

"Oh, this isn't a tiger." Lee put her arm around Pruno's neck. "See, He's very gentle." Pruno waved his tail.

The man, whose name we later learned was Mr. Ramaswami, wasn't interested.

"Please to remove tiger," he whimpered again. "I go out!"

Van pulled Pruno to the other side of the none too large car. Craig and Yorick followed. Lee and Christy arranged themselves across the back of the elevator and I, with Tucky draped over my arm, moved to leave space around the terrified man.

When the elevator stopped and the door opened, Mr. Ramaswami shot through the door like a stone from a catapult and before Van could push the "up" button, had disappeared down a corridor. The speed and agility of his rotund little body was astounding.

The kennel, when we reached it, was large and airy. Though a huge cage had been prepared for Pruno and smaller ones for Yorick and Tucky, after their caretaker met our pets he decided to let them have their beds on the floor of the large room. He showed us their own fenced, outdoor deck, a huge space the width of the ship and approximately thirty feet long. It was a deck above and overlooking the swimming pool. Our animals would have as pleasant a voyage as possible. We left them in caring and capable hands to search for our own accommodations.

Our cabins were comfortable and as soon as we had deposited our carry-on belongings, we went on deck for the departure ceremonies. Stewards carrying great baskets filled with rolls of serpentine paper offered each of us handfuls to throw at those poor souls left on the pier. A band played, but as the gangway was rolled away something hit me in the back, almost pushing me over the rail. Hot breath in my ear told me it was Pruno, and there he was, paws on my shoulders, tail wagging! There, also, was Yorick, happiness radiating from every tuft of black fur. Tucky was missing, but we prayed that he hadn't gotten off before the gangway was removed and that someplace on the vast ship a white cat was wandering alone.

The search was intensive and short. The cabin stewardess assigned to take care of us on the voyage found Tucky sitting outside our cabin door and brought him to the Purser's office. How did he know which room was ours? We'd been there such a short time and he'd started his hunt three decks above us. Was it intuition, smell, or just plain love that brought him to the proper door?

From then on, Tucky had the permission of the Captain to visit on any of the decks (on his leash, of course), and we had a long list of willing and eager cat-walkers among the children and adults.

The animal steward learned too, that unless you lock a gate securely, a Great Dane is as clever as Houdini.

The Purser told us that Mr. Ramaswami had arrived in his office, pale and panting, to report that a tiger was loose in the elevator and that he had only just escaped with his very life. Apparently the dogs had descended the only way they knew—by elevator, and with them, no doubt wedged into a corner, had been the corpulent and quivering mass of Mr. Ramaswami. That gentleman left our Purser in no doubt as to his feelings about the security of the ship and the safety of its passengers.

When we spied the Indian sitting with his round little wife in the cocktail lounge before dinner that first night, we hurried over to make our apologies.

Mr. Ramaswami saw us approaching and, abandoning his astonished wife, leaped up from the tiny table, brushed by us and disappeared through two sets of doors and onto the deck.

Mrs. Ramaswami, dressed in a lovely silver and blue sari, looked at us in bewilderment. "Where he go?" she asked.

Van bent over so he wasn't towering above her and said gently, "We have a large dog . . . ."

He got no farther before Mrs. Ramaswami began to giggle. "Scare he!" she said and covered her mouth with both palms. Her eyes danced as she rocked on the banquette, overcome with mirth. Van and I, not quite knowing at what, giggled with her.

"He say tiger in lift," she said at last. "You say dog.

Someday I meet dog, yes? Like be friend with somesing scare husband. Dog not tiger, no?"

We assured her that Pruno was not a tiger and that he was gentle. She insisted that she would like to begin her friendship with Pruno as soon as possible and left to pursue her cowardly husband. Before we parted we made an appointment for the next morning on the dog deck.

Mr. Ramaswami did not reappear until dinner time and then he was alone. Poor Mrs. Ramaswami! We guessed she had either taken him to task for being a coward or had laughed in his face.

She was, however, on time for our appointment the next morning. Though her first approach toward Pruno was somewhat timid, friendly Yorick won her heart and guided her toward his companion. She patted the Great Dane gingerly at first, but soon she was standing near the rail that overlooked the swimming pool, her chubby arm around Pruno's neck and her sari fluttering in the breeze.

The sun was warm, the sky cerulean blue, but the Pacific ocean was such a deep blue and so glassy it seemed made of ceramic. Only the ship's wake gave a hint that anything within our vision was moving.

All was peaceful until Mr. Ramaswami, his upper body welling out of too tight bathing trunks, sauntered by on the pool deck below. He was ogling the females in their bathing suits, as Mrs. Ramaswami no doubt had known he would. She broke into a paroxysm of artificial coughing.

When he looked up and saw his wife above him, her arms twined around the neck of a tiger, his expression was a fusion of terror, venom and incredulity. Poor Mrs. Ramaswami. I felt she was going to pay for her brief moment of superiority over her thoroughly detestable husband.

She looked at me with shining eyes, squared her shoulders and I had the feeling that for the first time in her married life, she had had the courage to do something on her own. Damn the torpedoes!

Day by day we understood Mrs. Ramaswami better. Every female between sixteen and eighty, except me, was pinched at least once by Mr. Ramaswami and some were pinched daily. If I hadn't been almost positive that the man thought I would sic the tiger on him if he included me in his painful play, I might have developed a complex.

After Mrs. Ramaswami bested him, he seldom brought his wife out in public, not even to the dining room. Our sympathy for the poor little woman grew and we were glad she'd had her amusement. We hoped she'd get the last laugh.

After a few days, perhaps in pique, Mr. Ramaswami formed an attachment for a wealthy, sixty-year-old widow from Fort Worth who, with her friend, Lola, sat at our table. Mr. Ramaswami may have been as old as she, it was hard to tell, but he seemed quite smitten.

Billie, the widow, did everything but stamp on his flat feet or hit him with an uppercut to discourage his attentions, but nothing availed.

Nightly he sent a bottle of champagne to the table, which she refused, and whenever she looked in his direction he bowed and twinkled as perspiration stood out and ran down his oily face. We all agreed that he was a despicable, unattractive little man. Billie declared him one of the vicissitudes of life and handled it all very well.

Mrs. Ramaswami, on the other hand, spent a lot of time on the dog deck when she was allowed free. We had the feeling that someday she would purchase a Great Dane of her own and that he would be brindle and striped like a tiger.

In the three weeks it took us to reach Hong Kong, Mr. Ramaswami neither spoke nor came near any member of the Vanderhoef family, human or canine. If Tucky was occupying the lap of someone near the swimming pool, which happened often, the man slithered by, his eyes on the cat instead of bathing-suit clad females. Everyone was grateful, and Tucky was always welcomed enthusiastically.

Every day aboard the *President Cleveland* was unalloyed pleasure. From San Francisco we swam and danced and ate and played our way to Hawaii, Tokyo, and on to Manila. We made wonderful friends who are still friends, and when we disembarked in Hong Kong, we disembarked as a body and moved to the Peninsula Hotel. Billie and Lola were there, but we thought we had seen the last of Mr. Ramaswami. Not so!

At mid-morning coffee in the hotel lobby, who should swagger by but Billie's bête noire?

But that's part of another story.

VIII

The day was glorious! Only a few airy clouds formed, then thinned and shredded and formed again to repeat their cycle in the infinite blueness of the sky. The calm water of the Hong Kong harbor reflected the blue; but surprisingly few sampans skittered from spot to spot on it, and only one or two red-sailed junks glided quietly through it. Tankers and freighters lay at anchor and the morning sun poured its light and a gentle warmth over the beautiful scene.

One of the freighters, the British-owned *Foo Chow*, would take us to Bangkok, and we were to meet the launch that would deliver us to her at eleven o'clock on the small dock two blocks from the Peninsula.

Van and Craig had to rescue our animals from quarantine and had engaged the same flat-bed truck they'd used to deliver the animals there. The girls and I gathered in front of the hotel to wave good-bye, but mainly to view the truck that had been the recipient, during our five days in Hong Kong, of very definitely condemnatory remarks by the two males in our family.

The large, rusted, flat-bed truck was just arriving when we emerged into the sunshine. It threaded its way around the semi-circular drive like a moth-eaten mastodon in a herd of sleek jaguars. Actually, there were some Jaguars lining its route. There were also Mercedes, and Cadillacs, and tiny MGs, to say nothing of one classic Duesenberg.

A fortune in expensive automobiles could have been wiped out if the Chinese driver of the truck had veered even slightly to his right or to his left. As it was, he sat, unimpressed, on a box (the seat having long ago disappeared to thieves, vandals or age), and peered myopically between the spokes of the steering wheel at the glittering array on either side of his huge, clattering behemoth.

The brakes and the clutch screeched as the driver writhed on his box to reach the pedals. The gear shift (minus its knob) protested, and the whole untidy mess came, surprisingly, to a halt.

Lee, Christy and I watched in amazement as the Peninsula's elegantly clad doorman produced a step stool and assisted Van and Craig aboard. He didn't even have to open a door. There wasn't one.

Again the gears screamed and they lurched away.

The girls and I decided to walk to our point of embarkation. We strolled along until we were passed by the truck, with Van and Craig teetering and skidding on the back of their unreliable conveyance, clutching leashes, plus cat carrier and trying to keep everything and each other from sliding off. The truck screeched to a halt at the dock.

When we arrived, six coolies and our mass of luggage were already occupying most of the available space on what had been loosely described as "the wharf." It consisted of a space possibly twelve feet long made up of boards running parallel to and bolted to the seawall that ran beside the road. It was no more than six feet wide and though it looked sturdy enough, I wasn't anxious to try it out.

The dogs had no misgivings. When they saw us they tore the leashes from Van and Craig and leaped from the truck, landing on the wharf within two feet of the luggage coolies. Of course the dogs flew to us, but not before one little man fled up the road and another backed to the edge and disappeared into the water. The other four coalesced into a clump and took root.

"Tiger! Tiger!" wheezed the only one who spoke English. The others reserved audible comment.

Yorick, loose from Craig's grip, bounded toward us, his black ears and red tongue flapping, his tail waving like a semaphore flag. Even when Lee had his leash in hand, Yorick continued to cavort and to cover us with sloppy dog kisses.

Pruno simply rose on his hind legs, put his paws on my shoulders and grinned with his face not an inch from mine. His tail waved with joy. I put my arms around him, hugged him to me and the coolies gasped in unison.

Tucky, in his carrying case, was already wearing his harness and leash, and when Pruno and Yorick were calm and the coolies seemed resigned to their fate, I took him out and cuddled him on my shoulder where he hung like a limp rabbit. The family was again complete.

As we'd walked to the dock, the red ferry boats had quietly come and gone from Kowloon to Hong Kong and back. Now, one approached and before it even came into the slip a low rumbling aboard drew our attention.

Suddenly, the hundreds of people who were crowded onto the ferry's decks raised their arms and voices and began to stream from the boat to shore. The rumbling became an ominous roar as the passengers pushed and shoved and jostled in what seemed a frantic effort to be freed from the confines of the ship.

"What do you suppose is going on?" Van set his briefcase carefully on the pavement and looked toward the ferry building. I was familiar with his watchful care of the briefcase.

Until we were aboard the *Foo Chow* and the briefcase was securely locked in the Captain's safe, I knew that it and its contents would be no more than a few inches from my husband's hand. The contents, all except one, were some classification of "secret"; that much I knew and that's all I wanted to know. However, in among the confidential information, was a snub nosed thirty-eight caliber revolver. I was only aware of that because British customs officers had required that it be placed in their custody during our stay in Hong Kong. Van had retrieved it when he and Craig had picked up our pets at the abattoir.

"It must be some kind of celebration," I said in answer to Van's question. "Didn't the waiter this morning say something about the tenth of October being the day of the 'Double Ten'? I wonder what he meant?" We both looked again at the ferry slip.

The din was growing louder and the men—they all seemed to be men—were still streaming from the boat.

They leaped or stumbled onto the wooden dock. They then surged through the ferry building and were suddenly joined by hundreds more who came from buildings, side streets and, seemingly, cracks, crevasses and holes in the ground.

We stood mesmerized. We watched as the grillwork of iron pipes that protected the ferry from interlopers was torn apart and the short pieces were raised above obviously overwrought heads. They became clubs!

In the minute or two before the seething mass of humanity, continuing to swell in volume and all armed with some manner of bat, turned to come toward us, we were more interested than frightened. However, I think Van and I recognized at the same time that this was more than a celebration of the "Double Ten"! The citizens of Hong Kong and Kowloon were inflamed, and we stood, suddenly trembling, in their path!

We faced a terrifying and powerful sea of humans determined to annihilate and we were as vulnerable as five tiny glass figures challenging a steam roller.

"Everybody move toward the water," Van barked and snatched his briefcase from the ground beside him. With it as a prod he began pushing at the kids all of whom were standing open-mouthed in fascination staring at the advancing rabble. Pruno's leash grew taut.

Craig, still with Yorick in tow, immediately got the message and pushing the girls ahead of him he moved as close to the water as he could get. The coolies gave way. Yorick strained to get back to Van and me, but Craig held firm.

My heart must have stopped in the first few seconds of recognition of danger, and when it began again it beat with such frenzy that it left me breathless and almost choking. Still, my mind worked and as I stepped in front of my children, Van stepped in front of us all. Our thin, little battle line was drawn.

Small hands fastened like talons on my bare arm and Tucky, draped on my shoulder, hissed at whatever was grabbing me. I stiffened and tried to pull away, but the claws held on and as I looked quickly I recognized the hands as belonging to one of the baggage coolies. The small man was trying to point out the launch which had left the *Foo Chow's* side and was racing toward us.

"Look Van!" I yelled to be heard above the cacophony. "The launch!"

Van's head turned. "It can't make it in time. When I yell, everybody hit the water!" I hoped we weren't going to be drowned like mice in a bucket, but Van had gauged the situation correctly. The water would give us a better chance than we had in the open. We could all swim. Even Tucky on his leash, as much as he might hate it, could swim. He would be given no choice. He and I would leap together. I knew Van had thought of his revolver; I'd seen him finger the catch on his briefcase, but it would have been of no help against the frenetic thousands who seethed and boiled as they advanced on our fragile line.

In front of us the mob raged closer!

I could hear a swelling murmur that meant the street behind us was filling with people and soon, when the Vanderhoef family was over-run, they would join the others to become a solid, raging, undisciplined legion.

The leaders were no more than fifty feet away now and I saw Pruno's ears flatten against his skull and his tail rise to became a straight extension of his spine. He stepped forward to the full extent of his leash, his ears came up and he stood like a perfect and beautiful statue in front of us all. His tawny coat shone in the sun and his stripes were as dark as if they'd been painted on with black enamel. He stood immobile, waiting.

The buildings on the left and the harbor on their right had compressed the frantic horde into a screaming, passionate marching army. They were close enough now for us to distinguish their features, but I saw only wild animals preparing to tear into their unlucky prey. Their eyes were angry and fixed, but as we watched and as the leaders neared, Pruno took his stand, and the expressions changed. Fear replaced anger; a small hesitation supplanted determination. The seething sea smoothly changed its course and surged around us.

The enemy, too, thought our beloved and gentle dog was a tiger.

The launch came in, the coolies shoved us and our belongings aboard, and we roared away, Pruno standing like a proud figurehead in the prow. Even Yorick didn't dispute his right.

Every foreigner who was on the streets of Hong Kong that day lost his or her life. There has never been any doubt that we owe ours to a gentle and gentlemanly giant with the soft name of Pruno.

IX

The dock in Bangkok, when we reached it after a long day of sliding silently up the Chao Phya River, was nothing like the bustling pier in *The King and I*.

It was crowded, but for the most part no one moved except the small men, presumably stevedores, who caught the hawsers thrown from the *Foo Chow* and made us fast to the cleats.

All the other humans, and there was a multitude, stood staring expectantly upward and when the chain was removed and the Vanderhoef family led by its hero, Pruno, started down the gangway, there was a subdued, harmonious "A-a-a-h!"

Rarely, since airplane use had become common, had a military family been crazy enough to come by ship, and certainly never had anyone brought a menagerie. The colonel who preceded Van, and his wife, had spread the word that oddities were docking today. Below us stood an audience composed of many nationalities, all curious, interested, perhaps even compassionate, but all ready to applaud or boo and hiss depending on our performance. Fortunately the actors, human and zoological, behaved with circumspection and were greeted warmly by the legitimate welcoming committee, made up of Van's predecessor, his wife and the four captains who would comprise my husband's command in Bangkok. All the others crowded forward to ignore us and to welcome the animals.

One of the most difficult parts of our two years in Bangkok started immediately with a "command performance" dinner party that night. In the weeks that followed, between luncheons, cocktail parties and dinners we tried to see our children and to find a house to live and entertain in for the duration.

Although we had a cottage on the grounds, the ancient hotel we had been registered in by Van's predecessor was not comfortable. House-hunting was uppermost in my mind.

Nothing was memorable about that two weeks except that at a luncheon, I met my first gibbon. He was small, black and furry with a white rim of fur around his bright-eyed, shiny black, pansy-shaped face. He walked entirely on his hind legs and, carrying the chain that led from a belt around his waist to a wire stretched between two trees, he ran to me and, with soft cooing noises, leapt into my arms and twined his arms around my neck. He had no tail, and, aside from his soft fur, might have been a human child.

My heart melted and I wanted a gibbon of my own! Van agreed that, as soon as we were settled in a house, I could get one. I stole even more time from the social activities to find a house as soon as possible.

It was new! The multi-colored, stone facade was shellacked and shiny. Balconies of pseudo-marble protruded from each of the three bedrooms and were surrounded by rococo, wrought-iron railings painted red. Downstairs the porches, that might have been obtrusive had they not nestled under the balconies, were pseudo-marble and had broad marble railings wide enough to sit upon. The stone, the marble and the iron were tied together by cream colored stucco. A portico overhung the driveway and provided shade for the porch. In the back a huge terrace, with a fast growing bougainvillea vine, was delightful for entertaining.

The house had few trees and shrubs for snakes to hide in (a danger in Bangkok where every tree and bush has at least one reptile), and it had, wonder of wonders, a bathtub and a small basement which, because it was built on delta land, was almost always wet. We rented the house!

There was no "better" section to the city and "nice" houses were scattered higgledy-piggledy among the opium dens, shops and crematoriums. Our house had a large, fenced compound with two good sized trees in the farthest corner and room for Pruno and Yorick to run. There was a nice-looking modern house with a red tiled roof behind us, an open field, with water buffalo grazing, across the lane beside us; and a ramshackle, unpainted wooden house on our other side.

The grass was tall and uncut around that gloomy-appearing abode. Trees overhung the house, and untended bushes all but obscured everything except a tattered blue and gray, striped canvas that seemed to hang over some sort of sleeping porch. The real estate agent assured me that the small canal between us plus the diligence of the gardener we would hire, would keep snakes from the area at bay.

We moved in, and within a day or two a little girl about Lee's age appeared at the fence that separated us from the disaster next door. How nice, I thought. Lee and Christy can have a Thai friend. Craig had already gone to a boy's camp down on the seashore where he could make friends and learn some of the Thai language. Now here was a chance for the girls to get to know someone other than another Caucasian.

Immediately, through sign language, I invited the child over to play. She indicated that she couldn't come, but invited Lee over there. Lee took her Ginny doll and its trunk full of clothes and trudged all the way around by Pattipat Road. I saw her disappear behind the tall grass and began to regret having let her go into what, I was sure, was a snake infested swamp. She reappeared briefly beside the spirit house, then was lost again for the afternoon.

All dwellings in Bangkok have "spirit houses" in their gardens. They range in size from small, twelve-inch or so unpainted houses with a porch-like affair on the front, to elegant three-foot-long, two-foot-high houses, gold leafed and painted to match the Temple of the Emerald Buddha. Spirit houses are designed to be put on a pole in the yard and, by providing a bowl of rice a day and burning joss sticks at all times, the house owner assures that the good spirits are appeased and the evil ones stay out of the main dwelling.

We had noticed that the owners of the house in which Lee was spending the afternoon fed their spirits an enormous bowl of rice twice a day, and kept a naked electric light bulb instead of joss sticks burning inside the spirit house.

"Did you have a good time?" I asked when Lee got home.

"I guess so," she answered as she plopped into a chair on the porch.

"What did you do?" I prodded.

"Well, not much."

"You must have done something for three hours."

The House in Bangkok.

"It was kind of boring. We just sat on the bed and Juey's sisters played all afternoon with my Ginny doll."

"Oh! How many sisters does she have?"

"Three."

"Did you meet her mother?"

"Maybe. There was a kind of old woman there, but I don't know whether it was Juey's mother or not. It's hard to talk when you don't speak the same language."

"Yes, I know." I thought about my trips with our driver, Bunchu. I always carried a pencil and notebook so that if I wanted to buy some flowers, I could draw a picture of what I wanted. If I wanted the pot, too, I could draw that as well. My drawings were rudimentary, but they got me where I wanted to go.

Juey came to our house a couple of times, but the visits weren't successful. She was into everything, closets, drawers and every cabinet that was closed. She was only curious, I'm sure, but Lee didn't enjoy her very much and soon the visits stopped.

Some months later, Craig and I were sitting on the porch watching Lee, Christy and some of their friends doing acrobatics on the trapeze hanging from one of the larger trees. I'd grown used to the samlors (pedicabs) and taxis that came frequently to the house next door, but a shrill laugh caught my attention and I looked over to see what was amusing. A girl, one of Juey's sisters I assumed, leaned out under the ragged canvas and tossed a mango to a man standing below her. He caught it, got into a samlor and left as I watched.

"What do you suppose they do over there?" I mused as the samlor pedaled away and a taxi arrived. The girl disappeared.

Craig looked at me with his mouth open and his eyes wide and disbelieving. "Don't you know?"

"No, do you?"

There was a long pause while my fifteen-year-old son searched for words. Finally he leaned forward in his chair, took my hand in both of his and said gently as if he were explaining to a small child, "Mother—that is no ladies' bridge club!"

Horror and rage filled me to the brim. How could I have been stupid enough to allow Lee to visit a brothel? How lucky that she had emerged unscathed. I was a danger to my children. No woman as naive and imbecilic as I should be allowed to be a mother. The rage passed, but the horror remained and I was careful from then on to view everything with suspicion.

Sometime later, many men, including a fire department with out-moded equipment, came, burned the house to the ground, cut the grass; and chased nineteen cobras, six banded kraits and an assortment of other poisonous snakes into our compound. It was lucky that on the advice of those who had lived in Bangkok longer than we, we had hired a Muslim gardener instead of a Buddhist. Buddhists won't kill anything, and I would have left Bangkok rather than share the compound with writhing, sinuous, reptiles. Thom, the gardener, was equal to the task and he dispatched them one by one.

There were still snakes in our trees and bushes; that, I knew, but Thom made a sweep each morning and I had to feel content that we were clear for the day.

We bought two mongooses, Pat and Mike, to help Thom out. They killed a few cobras before they left home permanently after also killing the neighbor's goslings.

Our neighbor suggested a goose, so we got Oddle Waddle. He did fine until he fell in love with the mother of the departed goslings and pined at the neighbor's back fence until we just donated him to the neighbor. Thereafter the neighbor showered us with gifts to build up his Karma.

We all watched with interest the building of a new house next door to replace the destroyed house. First the pile drivers came—two men with muscles who hefted a heavy log up and dropped it down on the piling. Once they had driven sufficient teak pilings into the delta mud to support the structure, the foundation was laid, and the building was framed out on the first floor. Every stick of wood was teak wood to thwart the white ants and termites, but the scaffolding (bamboo, tied together with raffia) trembled and wavered as the carpenters swarmed on it like monkeys. It never collapsed as we expected it would, and the house rose in record time. It was attractive, a nice couple moved into it, but it never quite had the interest and intrigue of the former shanty, or its occupants.

Yorick left the compound one day when we'd been in residence only a week or so. When he came home late in the afternoon he brought his new wife with him. We assumed she was pregnant and—she was. Fortunately she was mange-free, which was rare for the wild Thai dogs, but after the puppies were weaned and given away, she became vicious and we sent her to dog heaven. Yorick must have felt she had been a shrew and was happy to be rid of her, because he never made any attempt to find a replacement.

Craig came home from camp with a rudimentary knowledge of the Thai language, amoebic dysentery, and a host of Thai friends. He was invited to a Muslim wedding and when he came home afterwards he asked, as he walked through the door, "If I take the Kaopectate now, will it help?"

"What makes you think you'll need it?" his father asked.

"Well, when we got out of the car we walked under a sort of tent thing where there was an old man sitting cross legged on a water buffalo hide. He was carving the hide into a long strip. We all walked across the hide to get out the other side and I guess all of the other people did, too."

"Were you with or without shoes?"

"With."

"Then what happened?" I joined the inquisition.

"The bride and groom came out of a house and got tied together by the strip of buffalo hide when the man was through carving it. I guess that meant they were married."

"Did you have anything to eat?"

"That's why I need the Kaopectate. After Widnai and his bride wandered around for a while tied together, they took off the strip of hide, boiled it in a big kettle and then we ate it."

"You ate it! How did you eat it?" I was horrified.

"They had a conference, and since I was a falang (foreigner) they thought I should have a utensil, so they dipped a big sort of kitchen spoon in the dirty canal to wash it off, and gave it to me to eat with. They ate with their fingers."

"How did you cut it, or was it in pieces by then?" Ever the engineer, Van got to the crux of the situation.

"No, it was still in a long strip, but I scraped the hair off it with my pocket knife, chopped off a piece, ladled it in with the kitchen spoon and swallowed it whole. I thought if I chewed I might throw up."

I knew how he felt. The week before we had been the guests of honor at dinner in a Chinese home. There were eight of us, three Chinese couples and the Vanderhoefs. The round table was set in the middle of a large, airy dining room with a tile floor, and servants came and went on bare feet.

The first course was shark fin soup and, not having been indoctrinated into Chinese customs, since I felt I couldn't swallow the shark fins without regurgitating, I had left my bowl half full. That, it turned out, was the epitome of insults and I spent two years apologizing for it. Not only had I insulted the host and hostess, but they had only one bowl per person and the resulting confusion while they tried to equip me to continue the meal, kept us at the table an extra half-hour.

Course after course followed, each one more of a delicacy than the one before. By the end of the meal—and it did eventually come—every muscle and sinew was so stiff with apprehension that I could hardly close my fingers around the spoon or chopsticks. We had bunches of unborn baby birds tied together by their spindly legs, which everyone else munched with relish. As the bones and beaks cracked and crunched, the other guests "oohed" and "aahed." I can't describe the noise I wanted to make. A soup of chicken lights followed. Among the other horrors we had bird's nest soup, the nests having been made by the mother and father birds throwing up seaweed and weaving the results into a nest. Ours, the host was proud to point out, had been well used which, he said, added to the flavor. Tree fungus in sugar water proved an almost welcome respite.

Some courses, there were ten in all, I forgot as quickly as possible, but each was more revolting and harder to swallow than the one before.

Lee and Christy went to a Chinese wedding not long afterward and brought home knowledge that I would, gratefully, have used at the dinner party. When the wedding guests didn't care for the course in their one bowl, they emptied the bowl under the table, to be ready for the next course. Lee came home with her patent leather slippers full of shark fin soup. Her neighbor had had the same antipathy I did.

Some of our experiences in the two years we lived in Bangkok were onetime things and we remember them with pleasure and nostalgia, but the shark fin soup was a staple at every dinner party and only the forgiveness of my Thai and Chinese friends made it possible for me to become acceptable in their society. I never learned to swallow the fins.

In appreciation of my travail at the Chinese dinner and to reward myself for not disgracing us more than I had, the next day was the day I went to find my gibbon. My Penny!

X

In the late afternoon, spasmodic breezes often made the shaded front porch a pleasant place to regenerate. The days in Bangkok were hot and filled with required activities. In the evening there was nearly always a dinner party, a reception or some other official function. Most of these were classed as "command performances" and it was an unusual and joyous occasion when we could have dinner at home.

I tried to leave the afternoons free to spend time with the kids and Penny. Generally I was successful.

One afternoon, when Penny had been with us a month or two, we were sitting on the porch enjoying the breeze and the quiet. Even the traffic on Pattipat Road seemed hushed and farther away than usual.

Lee and Christy had borrowed a water buffalo from the field across the road and, having hosed her down and fed her green grass, were riding her around the compound followed by Yorick. Every so often the parade paused while the green grass was recycled. Pruno and Tucky lay on the cool marble floor of the porch, and Penny, since it was getting near her bedtime, was stretched out on her tummy on my lap.

Suddenly, the gate burst open and Craig, on his bicycle,

came hurtling through. Around his neck clung a tiny gibbon, its black eyes shining with excitement and fun, mouth open wide to catch the wind. He was half the size of Penny and was then and remained always, the epitome of the joy of living.

It was obvious to all that Craig had been drawn into the "pet shop," but from the moment he transferred the gibbon's baby arms from his neck to mine, saying, "Here, Mother, I brought you something," there was never a regret. Here was no frightened, pathetic creature as Penny had been. Pogo was bursting with happiness. He had a curiosity that knew no bounds and was always an enchanting, inventive, lovable imp.

He gave me a quick grin as he slid down my body to the chair seat and wriggled to the floor. It was all so fast that Penny, almost asleep on my lap, missed the performance. The little gibbon didn't see her either and I wondered what would happen when they became aware of each other.

Once on the floor, he flung himself on Tucky who had been resting quietly on the semi-cool, pseudo-marble. Poor, long suffering Tucky rose with great dignity and marched across the porch towing a gleeful six-inch gibbon at the end of his tail. The gibbon made a funny little sound almost like a giggle.

Penny's head came up and when it registered that here was another, truly of her species, she clambered down, threw her arms wide, rushed to the stranger and before he was willing to let go of Tucky's tail, had him enfolded in a vise-like, but tender embrace. The meeting dispelled any reservations we might have had at the prospect of having two gibbons. We cheered!

Lee, Christy, and two friends with water buffalo.

"What are you doing up there?" Lee called from the back of the bullock.

"We have a new gibbon," Craig answered as he dropped into a chair. "Come see!"

Christy immediately flew from her perch on the docile beast's broad back, leaving Lee to urge it toward the gate and the field beyond.

Craig's bicycle was leaning against the steps, and Christy, who always chose the hard way, tried to hurdle it instead of moving slightly left or right to miss it. Fortunately nothing about her either split or broke as she hit the porch with a splat, but we all rushed to her in apprehension. We had finally bought stock in Johnson and Johnson hoping to get some money back from the Band-Aids always plastered on our youngest.

The new gibbon tore himself from Penny's grip and followed closely by his mentor, rushed to the prone body and flung himself on Christy's head. Penny followed suit and under the tangle of hair and gibbons, Christy began to giggle.

"I'm all right," she managed to say, "but they're pulling my hair!"

Lee had come up the steps, and she gently lifted the littlest gibbon from her sister. I disentangled Penny, Christy bounced up and we all moved to the other end of the porch and sat back down. Penny struggled to get to her new brother and when I let her go, she rushed to Lee's lap and again enfolded the unwilling baby in her arms.

"Look, Mother." Lee detached the gibbon and held him out so I could view him. "The shape of his head is just like Pogo in the funnies. Let's name him Pogo."

"I think he looks like Joe E. Brown. Look at that long upper lip and wide mouth. I vote for Joey!" Craig leaned back in his chair.

"I think we ought to name him that word I learned at lunch. What was it?" Christy straddled the broad porch railing and bounced up and down as if she were still riding the buffalo.

"The word was 'grandiose'," I said, "and it means pompous and showy. Does that little furry thing look pompous or showy to you?"

"I don't even know what 'pompous' means. How could I know if he looks like it?" Christy was indignant.

"Believe us," Craig said, "he doesn't, and don't ask us what it means. You won't remember it if we tell you."

"Let's vote on his name." Lee set the un-named gibbon on the floor.

There were two votes for Pogo, one for Joey and one for Pompom, Christy's version of "pompous." "Grandiose" had disappeared from her memory. Pogo had his name.

When Lee put Pogo down we all watched with fatuous

smiles as Penny made an effort to introduce her new companion to people and animals alike. When Pogo first stood, we realized he was such a baby that he was just learning to toddle and that his progress was as unsteady and full of bumps as any human baby's.

Penny tried to be patient with his ineptitude, but she gave up and dragged him along on his back, towing him by one spindly arm. Pogo thought that was fun and grinned widely as she pulled him to my feet. Penny showed no jealously at all as I lifted Pogo onto my lap, which surprised me; she climbed up and sat close, watching with as much pride as if he were a product of her own conjuring. Penny was definitely the motherly type.

Trying to examine Pogo for injuries was like trying to examine perpetual motion. I found, to my pleasure, that he was un-injured. He had a sturdy little body and his pansy-shaped face surrounded by white fur was rounder than Penny's, and he also had a much longer upper lip. He did look like Joe E. Brown, and as the examination continued he grinned at me with the same wide, friendly grin.

Everything delighted him and his eyes snapped with devilment as he investigated me. I must have passed inspection, for he climbed up, put his arms around my neck, cooed, and slithered to the floor where he made his unsteady rounds like a slightly tipsy, cocktail party host. Penny stayed with me and climbed onto my lap as soon as it was vacant. It was nearing her bedtime and I wondered what would happen when five o'clock came.

"Here, Lee," I said and tossed her a small banana from Penny's supper plate, "give Pogo this. He must be hungry."

Pogo took two bites of the banana, dropped it on the porch and pounced on a napkin that had blown from the rattan table.

"Look at that," Craig laughed as Pogo put his forearms on the napkin and straightened his legs to raise his bottom in the air in an upside down "V." He began to push himself with his legs around the slippery floor. When he was moving at a good clip, he did a belly-whopper on the napkin like a child on a sled, and slid a foot or two. He, and we, found it most entertaining. He repeated his trick until he ran into the wall. Without even taking time to think about it, he grabbed the napkin, threw it over his head so he looked like a tiny ghost, and, with outstretched arms, staggered blindly around until he was again floored by the wall.

Once or twice the napkin fell off and we could see that his eyes were tightly closed, but he quickly draped himself again and went on with his game.

Promptly at five, Penny got down and with her graceful gait, went inside and, I presumed, up the stairs to bed. I think she expected Pogo to follow, but he was still more interested in sliding around the floor. After a few minutes the screen door opened and Penny peered out. She whooped at Pogo who sat up to listen, but when she disappeared once more, he went back to his play.

The next time her worried little face peeked out, one of us, rather than have her upset, scooped up Pogo, and we all trooped up the stairs to see how bedtime went.

The trip, as did everything else, proved stimulating to Pogo. In the dressing room, he was wide awake and ready to recommence his activities. Penny scrambled straight up to her shelf and grew furious with Pogo, who hadn't yet learned to climb, for his inability to follow. She was like a bossy little girl and made several trips up and down urging him on. Before she grew completely frustrated, and though he seemed far from ready for sleep, I lifted Pogo onto the shelf with her.

They each had a banana for the night in hand, and that was a mistake. Pogo used his like a bat and flailed the air and Penny with it as she flattened him onto his back. She dropped her banana, squashed Pogo with one hand while she haphazardly adjusted the doll blanket over both of them. As the four of us watched, it appeared to he a losing battle for Penny. No sooner was everything arranged to her satisfaction than Pogo would be up and undoing.

Finally, Penny made some statement in gibbon language that caused her reluctant companion to stay put and to close his eyes. Both seemed to be asleep in seconds. As we left the dressing room I looked back and thought I glimpsed one bright, beady, impish eye peeping from under the yellow doll blanket. I wondered whether Pogo would waken Penny when he got up again or whether she would sleep blissfully on, never knowing she had lost the battle.

For her sake and ours, since Van and I had to sleep in the adjoining room, from the bottom of my heart I hoped Pogo would, at last, succumb.

XI

One morning, two or three days after Pogo came to live with us, we wakened to find that the temperature had dropped in the night and we were, of all things, cold. We were used to a minimum of ninety-eight degrees, with humidity almost to match, and sixty-seven, and little or no humidity, might as well have been below zero. We rushed around unpacking long-stored sweaters for ourselves and the servants, but we had nothing for Penny and Pogo who were shivering on the closet shelf.

Gibbons are highly susceptible to pneumonia which we now know is a virus. In 1956 we thought if you got cold you could get pneumonia, so it was essential we find something to put on our sparsely furred babies. Finally, Ampon, our new "Number One" servant, cut up one of the doll blankets and made two, tiny, yellow jackets that covered the gibbons' chests. They pinned in the back with safety pins. We had no time to devote to niceties like buttons or snaps. There was a little excess material in the rear which, when bunched together and pinned, stuck out like a bustle. It took the garments out of the ordinary. They looked cute.

The gibbons didn't object to being dressed; in fact, Penny sat in front of the mirror on my dressing table admiring her reflection and smoothing the material over her chest and tummy until she almost missed the exodus to breakfast.

The sun tried to warm the house, but outside it only reached seventy all that day. Pruno was ecstatic. After months of lying on the marble porch, panting in the shade, he cavorted and romped in the compound with Yorick, more like a puppy than the eleven-year-old he was.

Tucky, who enjoyed warmth, curled into a tight hall and we covered him with another doll blanket. I'm not sure he moved for the three days of winter the rest of us found so life-saving. Since the temperature drop never happened again while we were in Thailand, it obviously was a fluke, one the Americans would have liked to have had repeated. The Thai were not as enthusiastic.

When the sun reached the back terrace on that first cold day, and after Bunchu had driven the kids to school, I took the gibbons out to play in the bougainvillea vine. The terrace was large and square with a tile floor, and it was elevated three steps above the ground level. Instead of a railing, the terrace was surrounded by highly polished slabs of pseudeo-marble, each about twelve inches wide, three inches thick and five feet long. The slabs were attractive and made extra seats for any number at a party, providing the guests were liberally doused in insect repellent for the mosquitoes. Trellises with bougainvillea in a variety of colors shaded the terrace in places, and made it a lovely spot to sit.

Thom had already made a thorough check for snakes in the vines and the shrubbery that surrounded the terrace. That was an absolute necessity in Bangkok where every bush, tree and shrub had at least one slithering resident. Many were poisonous, a few were not.

I ignored the shade and pulled the table and chairs out

into the warm sun. Penny, in her yellow jacket with the peplum, slid from my lap and quickly danced over to the nearest vine. In a second she was up among the flowers looking for insects. Pogo, never having been free outside since he had been snatched from his mother, took a few minutes to look around and to make his plans.

He sat on my lap viewing the wonders of his world, and when he was ready, he went down the rattan chair leg as if it were a fireman's pole. He was much more adept at swinging than at walking on two legs, but he toddled unsteadily straight to the edge of the terrace and down the steps. Thinking that he might be planning to run away, I followed, but he only went as far as a plant that had leaves like a geranium and a tall, pliable stalk, maybe four feet high, shooting from its center.

Hand over hand and foot over foot, Pogo climbed the stalk until, like a sapling in a wind storm, it bent and bounced him up and down. When it became still, he dropped to the ground and repeated his climb time after time. The funny little "eeeee" noise he made as he bounced up and down, and his wide grin, brought pure pleasure.

I called Ampon to watch and the two of us sat like fatuous grandmothers sharing his joy.

"Must go do work," Ampon said at last when Pogo left his elastic plant and climbed back to the terrace. "I bring Madame iced tea." She went inside and closed the screen door softly.

By the time she returned with the tea and a banana for the gibbons, Penny had left the bougainvillea vine and was swinging from the heavy teakwood shutters that were our only windows aside from the screens, and was searching behind them for spiders and other insects

Pogo and his "bendy" bush.

Craig and Penny have a chat.

Pogo, too tiny to reach the bottom of the shutters, sat on the tiles below Penny and whooped mournfully. Ampon set the tray down and lifted the baby.

Once one tiny hand and a foot found a hold, Pogo went up the shutter like a fast moving paint brush. He sat down on top, his bottom and feet firmly planted on the thick window cover, knees under his chin and arms wrapped around his legs. His yellow bustle stuck out in back like a roostertail, and whatever sound he made as the equivalent of a cock-crow said, "Hey, everybody, look at me!"

Ampon, Penny and I looked. So commanding was Pogo's whoop that even Pruno and Yorick raised their heads, and Thom came hurrying around the corner in response. Satisfied, the gibbon slid off the top and picked himself a handful of bugs from behind the shutter. These he ate, then he slid down and plopped to the tile floor.

"Look, Madame," Ampon said as Pogo stood again at the top of the steps. "He pletend he not see zat frower before."

I don't know how that little gibbon signaled his thoughts to us, but there was no doubt as to what he was doing. His thoughts must have been, "Look at that. Where do you suppose it came from? Lawsy me! Let me at it." And so down the steps he went again to begin his climbing and bouncing routine. This time he embellished the game by playing it with his eyes closed.

That plant withstood Pogo for more than a week before the stalk broke, but by then Thom had a replacement ready and if Pogo ever knew the difference, he didn't indicate it. It was always his favorite jungle gym when we were on the back terrace and even when he grew larger and it whipped him up and down until he often hit the ground on his back, he spent hours on it. When we had accumulated quite a grove of the things, Pogo expanded the game to include staggering from one plant to the other with eyes closed and arms out-stretched. This was a variation of his game of trying to walk with something over his head. Once he tried to combine the two games, but the rag he used got wrapped around the sticky stalk and frustration set in when he couldn't get it loose. Showing his ability to reason, he never tried this again.

The gibbons had, of course, a basket full of toys that stayed beside a chair in the den. Penny liked the soft, cuddly things, but the harder and wilder they were, the more Pogo enjoyed them.

His favorite toy was a string of four, hard-rubber ducks about two and a half inches high. They were separated by flexible rods and each duck had small wooden wheels. The toy was obviously meant to be pulled, but Pogo had his own game. He'd grab hold of the string attached to the first duck and whip it back and forth until he got it up to his required speed, and then he let go. After the toy flew through the air (sometimes across the room and sometimes no more than two feet) and fell to the floor, Pogo rushed and flung himself upon it, and they wrestled like two living creatures. Since the toy was flexible, it gave him quite a tussle.

Once, in the breezeway between the house and the kitchen, he met a long, green, tree snake. With every intention of wrestling it as he did his ducks, Pogo advanced upon it. It struck and missed. Pogo advanced again. I, of course, having a phobia about legless reptiles and unable to even view one in a photograph, screamed for Thom. I really wanted to flee, but terrified for Pogo, I stood rooted to the porch above the approaching mayhem.

Thom was right around the corner of the house, and as the snake readied itself to strike again, he hit it with a shovel and it demised before Pogo was bitten. The snake was a deadly poisonous one.

"Watch out," my son said with great wisdom. "This thing will have a mate someplace around here."

A week later, Craig was leaning against the back kitchen door that opened out onto a small canal. He was shooting rats, mortal enemies of all Americans in Bangkok, and the source, I presume, of the snake's meals. He had his air rifle to his shoulder and was taking aim when the predicted mate dropped from the door top and wound itself around the gun barrel. Craig threw the gun, snake and all, into the canal, and slammed the door. Again, a fearless Thom was called to arms and went in to check for further relatives. The gun is, no doubt, still in the klong today.

Because of the mid-day sun in Bangkok (made famous in the British song, "Mad Dogs and Englishmen Go Out In the Noon Day Sun"), we all napped after lunch. So that we could have quiet, the gibbons were separated. Penny slept with me and Pogo went with Lee and Christy.

The girls had an air conditioner in their room, but in the daytime they preferred the oscillating fan. The first day Pogo napped with them, Lee came tiptoeing into my room.

"Come look, Mother." She had a smile on her face and her hazel eyes were twinkling.

"Is it worth it?" I asked.

"I think you'd be sorry if you missed it!"

Penny was sound asleep so I got up quietly and went across the wide hall that was furnished as a sitting room.

The fan was blowing across the twin beds, and seated right in front of and about a foot from it, was Pogo. His Joe E. Brown mouth was wide open so the air could blow in, his eyes were closed and he was sound asleep. I reached out and pushed him over gently so he was lying on the bed. He didn't waken nor did he close his mouth.

Every day he started his nap sitting in front of the fan and one of the girls gave him a push after he'd fallen asleep. Gibbons perspire just as people do and I expect the fan felt awfully good to his little, hot, furry body. By opening his mouth, he seemed to be trying to cool himself from the inside out.

Nap time was the only time I can say that Pogo was quiet. He was an imp, and devilment shone from his eyes. Intelligent and inventive, he and Penny, who was just as smart and inventive, created havoc at least once a day.

Sugar, whether it's natural or refined, causes diarrhea in gibbons, so our sugar (plus all of our fruits), was locked in a screened cabinet in the pantry. The cabinet had three shelves and was shaped like an old fashioned pie safe. Each of its four legs stood in a bowl of water. to keep the enormous, two-inch-long roaches from getting at the foods stored inside.

When Penny first came, before we learned much about gibbons, the cupboard was unlocked. One day, when she either smelled or saw the enticing contents of the safe, she climbed up and swung on the door knob with such force that, unsteady on its legs in the bowls, the safe fell over on her.

I heard the crash from the living room and rushed in, terrified that the little gibbon must have been crushed.

When the cabinet was lifted, Penny was sitting between two shelves, the shards of the broken sugar bowl around her and white granules streaming from both hands as she shoveled the delicious, forbidden food into her mouth. Fortunately, we had a new bottle of Pepto Bismol. We should have bought stock in the company right then. We used a lot in the six years ahead.

Naturally, we then put a lock on the cupboard. Penny mastered it. We got a new and different one that required a key. That worked until someone left the key on the counter. Penny was no dummy. She watched carefully and the next time the pantry, where the cabinet sat, was empty of humans, she tried and succeeded. We bought another bottle of Pepto Bismol.

Penny and Pogo with author.

When Pogo came we acquired a hook latch with a spring that took two hands and strength to open. Opening it became possible when two devilish minds and twenty nimble fingers worked together. Lots of Pepto Bismol! No shortage of thought and reasoning from the gibbons either.

Finally, we sealed the sugar and fruit in the padlocked

refrigerator on the back porch and the gibbons never knew where their treats had gone.

Another of Pogo's games and one that filled him with demonic glee was to swing behind me outdoors, or run on the back of my chair or the couch, grabbing a handful of hair in passing. He was so fast that I never had time to brace myself and I lived with a whiplash all the time we had him. It wasn't a major whiplash, but my neck was never completely comfortable. He didn't assault anyone else and if I hadn't recognized the devilish gleam in his shoe button eye, I might have thought the game was a sign of affection. The glitter was there, however and the added "eeeee" and grin were a give away, so, because I loved him and because I couldn't think of any discipline that would stop him, I tried to keep my back protected, like Wild Bill Hickok. Also like Wild Bill, I was unsuccessful.

Sometimes, after one of his fiendish ambushes, he would bring his prized possession, his towel, and drape it haphazardly on my head. He never left it very long before it was over his own head and he was staggering blindly around on the floor, but the ritual was meant to say, "I'm sorry" and to show me that he did love me after all. He had many ways of getting to the bottom of my heart and that was just one of them. I miss him now and I expect I will always miss him.

XII

George was a big, black, rowdy gibbon. If he'd been cast in an early "Western," the director would certainly have chosen him to play a Charles Bronson part and he certainly would have portrayed an outlaw. Actually slim, as all gibbons are, his body gave the impression of burliness, and though he was only two feet tall, when he raced toward us at his flat-out speed we bunched our muscles and prepared to take possible flight.

In his "Western" he would have swaggered into a saloon, and when he'd had enough pleasure and too much to drink, he would have sashayed out leaving the premises in a shambles and the other patrons lying exhausted on the floor. He could have done it! He did it to us every time he unhooked his chain from the ring attached to his owner's house across the lane.

Our compound and house were George's saloon. Each time he mastered the new catch (his owners kept trying to control him) we were besieged with little warning.

He came bursting through our screened door, his chain draped over his arm like a bride's train, and every one of us hurried to get outside while the inside of the house was still intact.

The first time George appeared unexpectedly and unescorted, he flung the door open; it banged flat against the wall behind it, then rebounded and slammed with such might that the handle fell off.

Before we could get up from the lunch table, George was up the stairs, through the decorative openings between the stairway and dining room and, from the swinging chandelier, had dropped down into the middle of our lunch. Plates, glasses and food flew as he raced upright the length of the table and launched himself at Penny who was sitting beside me in her doll high-chair.

Craig and Van had gone for a swim and lunch at the Sport's Club, so only Lee, Christy and I were at home that Saturday. When the black fury demolished our lunch we fled from the dining room in terror. George's arrival was so cataclysmic that I don't think any of us knew at first who or what he was.

Penny's high-chair, of course, fell over with a crash, but Penny, as agile as George though less than half as big, had recognized a playmate and was already out of the chair.

Immediately, the two began chasing each other rapidly around, under and over the table, swinging from chair back to chair back. They raced past us into the living room and having toppled the floor lamps and swept all end tables clean, they raced back.

Pogo, the eternal imp, temporarily ignored the mayhem and taking advantage of the lack of supervision, seated himself beside the plate of fresh pineapple. Loaded with natural sugar, the ripe and delicious fruit was a sure-fire diarrhea producer in gibbons. Before I could steel myself to perform my motherly duty and remove Pogo from temptation, the plate flew to the floor and largesse was scattered the length and breadth of the room.

In the archway between the dining room and entrance, huddled beside the still quivering screen door, we humans repossessed our mentality and I, as the senior member, became aware of my duty. I reverted to a, more or less, responsible mother, and knew there had to be a salvage operation!

"I'll grab George," I said although I was somewhat afraid of the big brute, "and the rest of you bring Penny and Pogo. Let's get them outside." The salvage team waded into the scene of destruction.

As the gibbons had writhed and wrestled joyfully in the middle of the table, George's chain behaving like a noisy, independent, thrashing reptile, plates and food had been pushed to the floor. The dogs, always banished from the dining room at meal time, assumed that lunch was over. They bounced in and, ignoring the thumping and clattering above them, cleaned the rug of any ham or egg salad sandwich that might have fallen. They did leave the limp lettuce leaves on the rug, a pale green pattern on the light tan sisal.

The peeled slices of pineapple were wrenched from Pogo by the tumbling apes and landed with juicy plops, ignored until I stepped on one. Like an ice skater, I glided giddily across the sisal rug and came to a stop only when I hit the table. When my femur recoiled from the table's edge it was natural to bend, and as I did, my out-stretched hand closed upon George's long fingered black one. With my other hand, I grabbed the end of his chain and when he tore his hand loose, I still had a firm hold on him. Because George showed no animosity as I tugged him from the table, I think he was really delighted to have someone else carry the burden of his chain. This freedom allowed him to grab a slice of pineapple in each hand as he was rushed past. Since, by then, the fruit was all over the room, this required no special effort.

We swooped through the door and out onto the porch

before he even had time to take a bite. Lee and Christy brought out Penny and Pogo and we hurried to the mimosa trees where we had a table and some chairs.

It wasn't long before our two small gibbons were exhausted by the rough play. After all, they were still babies but George was fully grown, and even after Penny and Pogo had crawled into our laps for protection, George went right on climbing, swinging and running. He made several attempts to pull Penny into his games again, but when she wouldn't join him, he went off and shredded our six banana trees.

Finally it grew quiet and I hoped he had gone home, but when I heard a scream from Molli, our wash ayah, I knew he was still with us.

George, bored with playing alone and having been relatively good for almost an hour, had invaded Molli's territory and bitten her.

The cook came running.

"Madame," shrieked Tong Sook hustling toward me, toes splayed out and waddling like a duck. "Madame, do somesing. Molli be bited."

I leaped to my feet. Visions of a ravaged Molli sent me racing toward the building that housed the servants' quarters, laundry and kitchen. Penny and Pogo clung around my neck and emitted soft whoops as I ran, trailed by Lee and Christy, Tong Sook and the dogs.

Molli, with her basket full of wet clothes was barricaded in Tong Sook's room but we could see George crouched on the roof of the house behind us, tearing the terra cotta tiles from it and hurling them at the gardener below. Our own gardener, Thom, looked on through the fence in amazement.

"Molli," I said through the door, "let me in. I want to see where you're bitten."

"Bite bad," Tong Sook offered.

The door opened a crack and Molli's pretty face, unscarred as far as I could see, peered out. When she was sure that George was not in sight, she let us in. I looked for flowing blood; there was none. High on her bare arm near the left shoulder, was the bite.

It had hurt, I'm sure, and probably was still stinging somewhat, but by no conceivable rule could it be called a bad bite. George had long, sharp eye-teeth and could have done as much damage as a German Shepherd, but this, fortunately, was only a nip. I began then to think that George was a big fake who got his laughs out of terrifying and dominating those humans with dark skin who had snatched him from his free life and chained him to a house. In the two years we knew him, I never heard of his biting anyone whose skin was white.

Still, Molli was hurt and afraid, and it was my job to see that she and the other servants could do their work without fear. With Penny and Pogo still clinging to me, I went back around the house and out the gate. The entourage followed.

Across the dirt lane, the American captain, George's present owner, was just setting out on the run to rescue the Thai neighbor's roof and gardener. No one in a mile radius could have ignored the yells of the beleaguered gardener who was being hit by the tiles George threw. Nor could they have disregarded the frantic screams of the house owner. Standing in his doorway, well protected from George's missiles and watching more and more of his tiles become shards on the hard ground, he hurled vituperation at the hapless employee who, afraid of his master and terrified of George, didn't dare seek cover.

When Allen, the gibbon's owner, appeared in George's line of vision, the ape grinned, left his pitching practice and swung down into Allen's arms. The apoplectic Thai owner stopped yelling at the gardener, who scuttled around the house and disappeared, and turned his invectives on Allen. Since the Thai man spoke no English and Allen spoke no Thai and none of us wanted to translate, the noise grew less and less and ended in much bowing. Tong Sook did translate Allen's apologies and his offer to pay for the damaged

George with his owner. Yorick joins the fun.

and missing tiles, and from then on, the gathering was down-right cozy.

Allen bought a different kind of latch for the chain and we had two days of peace before the gibbon figured out how to work it.

Once again we heard the clatter of tiles, the scream of the neighbor's gardener; and then the screen door burst open and the roundelay began. We learned to prevent it. The pattern was set. Whenever George came, and he came as often as he could and as soon as he could figure out any new latch, Thom ran out the gate, across the lane and brought back a member of the family that owned George. He would be taken home. Or, if he stayed a while to play with Penny and Pogo, and he did often, the family member stayed with him.

When supervised, George was not a nuisance, but was a clown and a pleasure, and we missed him when his owners went home to the United States and the gibbon was given to a family who lived farther away.

Once after he was gone, a big black gibbon appeared on top of our bougainvillea arbor. We thought it was George and went out happily to greet him, but that gibbon had a different face and was in no way gentle so we fled to the house and locked the doors. Thom said the gibbon lived at the crematorium down the road, and he rushed to alert the gibbon's owners. A Buddhist priest, in his saffron colored robe, came late in the afternoon and collected him and the gibbon never came again. Unfortunately, neither did George.

XIII

Wildlife additions to the household, if you happened to be a family of animal lovers, seemed a natural sort of thing in Bangkok. The weather was always sunny, the compound was large and there were servants to help care for and pick up after the animals. One by one we added to our menagerie.

To begin with, we had our two dogs and cat, of course, and soon Lee and Christy had rescued two guinea pigs from starvation. Craig, added a pair of mongooses. I had the two tiny apes who took up more time than the children, but Van still had nothing of his own.

He had refused the gift of a baby elephant, not because he didn't want it; he did, and the family urged him to take it, but when Van heard that the elephant ate half a ton of hay a day, he declared an elephant impractical. We had no storage shed.

When Van came home one day and announced that, in front of the so-called pet shop, he had seen a crane stalking about on the other side of the canal and that he was much taken with it, the family rushed out and leaped into the car. Van bought his crane and named him—what else—Ichabod.

On the way home, Ichabod sat in the back seat with the children. His long legs were folded forward on the seat; his gray feathers, bordered in white and edged in black, were smooth and unruffled. He seemed a friendly fellow and appeared to enjoy the car ride. He looked with interest at the passing sights until, with a bright and beady glance at Lee, he threw up four whole, dead fish into her lap. It was as effortless and silent as a smile, and Ichabod who, we later learned, orchestrated any situation in which he was involved, quieted the ensuing pandemonium. He spread his two-foot wings and held everyone in the back seat motionless.

Apparently, he felt better after his temporary motion sickness, as he next, still with wings outspread, shook himself vigorously and deposited a large, white blob on the leather upholstery. After that, he folded his wings, closed his eyes and went to sleep.

Released from the wings, Craig scooped the fish from his sister's lap and added them to the litter on the roadside. Kleenex handled the blob, but as usual . . . our car began to smell.

At home, inside the compound, the crane refused to be helped from the car. He clattered his eighteen-inch-long bill like a rhythmic set of castanets and extended his wings once more. We all backed obligingly away.

"Here," Van said firmly, "we can't have dissension in the ranks." "Come, Ichabod," and he reached into the car. The clacking bill tore his shirt and left a long red scrape over his ribs. Army discipline broke down.

While we all (the servants had joined us by then) stood expectantly at attention, Ichabod hopped regally from the car. His long yellow legs moving in slow motion, he stalked about and surveyed the yard as if he had always owned it.

Quiet, gentle Pruno gave the new inhabitant a cursory glance, but Yorick was an avid bird chaser and this was more than he had ever dared hope. He dashed at the crane who, as the dog advanced, arranged himself in full battle pose and, with beady eyes gleaming, clacked his bill only once. He drew blood. Yorick retired in surprise, pain and shame, and although he later sometimes chased Ichabod, he never again came within range of the bill. They both came to regard these chases as a game.

Now, when the bird had established his dominance, he resumed the majestic examination of his domain. A large flower pot of variegated, broad leaf grass sat on the railing of the breakfast porch and, as it was close to evening, Ichabod chose this as his sleeping place. With his long-toed, flexible feet, he mashed this grass into a nest and, ignoring the audience, closed his eyes and drifted off. After that, the gardener, Thom, kept three pots in rotation so that, as the bird stamped and matted the contents of one beyond recovery, another pot was ready as a replacement.

We had no canal, or klong where Ichabod could wade, but behind the house, the ground under the clothesline had been becoming increasingly muddy. Since there had been no rain, we were curious as to the origin of the water, but before we could make any determination, Ichabod moved into this swampy area.

He stamped about in the black, delta ooze, letting it squish between his toes, and if he was disappointed that it had no fish in the slimy depths, it didn't show. He appeared euphoric.

Molli, the wash ayah, was not as enthusiastic. Whenever she hung the clean, wet clothes on the line, Ichabod tore them down and trampled them into the mud. When she tried to retrieve them, he chased her. Thom always ran to carry the bird to the farthest corner of the compound, but it seemed simpler to move the line away from the crane's claim. On my insistence, Thom did.

When the muddy yard began to develop a smell so nasty and over-powering that we could hardly eat or sleep, we suspected the source of the water and sent for the owner of the house.

Mr. Paitoon came, stood on the concrete between the house and the kitchen, looked pained and said, "Not understand."

"What don't you understand?" I asked.

"Order tank from United States." Mr. Paitoon had a son who was studying architecture in California. "Not should smell. Retter from son say."

I looked at the perplexed owner. In his khaki British shorts and bush jacket, with a pith helmet squarely on his round head and a riding crop under his arm, he looked like a Buddha dressed for a masquerade.

"But it does smell, Nai Paitoon," I pointed out, wishing Van were at home. "I think you're going to have to dig up the tank."

The Buddha considered Ichabod squishing around in the swamp. "Perhaps bird break," he declared flatly.

"No, Mr. Paitoon, the bird did not break it. It was broken before he came. You are going to have to dig it up and fix it."

The rotund little man sighed, "Okey dokey." The slang expression did not fit with his haughty demeanor and I tried not to giggle. "Man come tomorrow."

"Please, Nai Paitoon, man come today!" I said firmly. I would not be intimidated.

Our landlord bowed with great formality, muttered, "Today," and marched around the corner of the house and out of my vision. Thai men did not like to conduct business with women.

"Thank you, Nai Paitoon," I called after him. That afternoon two men came and dug up the tank. We saw that when the septic tank had been installed, the workers had simply buried it without connecting it to anything. Since they hadn't known what it was for and no directions had come with it, the pipes from the bathrooms ended four feet short of the concrete box. Van was at home by then, and he explained to the men how the tank should be connected. They labored for three days, most of the time spent flirting with Molli or avoiding Ichabod, but at last they re-covered it with the same wet, smelly earth, went away and we had no further trouble with that part of the bathroom.

Of course, when the servants gave me six buckets of hot water for my birthday and I tried for the first time to use the bathtub instead of the shower, the men had to come again. The bathtub drain connected to a pipe that ended in the pantry sink on the floor below, and the geyser flooded the dining room as well as the pantry itself.

When the mud in the yard dried, Ichabod was desolate, and in recompense, we bought him a huge tin tub, filled it with water and mud, and twice a day, Tong Sook, the cook, tossed in ten or twelve small fish purchased at the local market. The bird, as he squished around in his pool, feeling in the mud with his long toes for the fish, seemed satisfied.

Van loved and admired his pet and when my husband came home in the evenings, Ichabod rushed to greet him with his wings outstretched. He never gave any threats of nipping Van, as he did to the rest of us. It was true mutual affection.

"Too bad he can't come in," Craig said one afternoon as Ichabod stood peering wistfully through the screened door.

Ichabod on the back terrace.

"Well, he can't. He isn't housebroken and his droppings are messy." I had watched Thom hosing the concrete outside, and was not ready to let the gardener turn the hose inside the house.

"Maybe I can fix it." Craig rushed out to the kitchen. In a few minutes he came through the front door, Ichabod under his arm. A string around the bird's body held a coffee can under its rear, but as soon as Craig and Ichabod reached the sisal rug, Ichabod, always prone to motion sickness, threw up his recent meal of whole fish onto it. Craig set his burden down, and started for the cleaning materials....

"Watch it!" I yelled, remembering the chronology of the trip home in the car.

"The coffee can will take care of it," my son said confidently.

The coffee can swung as Ichabod stalked. He missed it.

"Out," I pointed to the door. "Out right now!" They left quickly.

Craig and I came to an agreement. Ichabod was, and would remain, an outdoor pet.

The bird was useful as well as decorative. He and I shared a phobia about snakes, but he was able to do something about it. His long bill was lethal to the poisonous and non-poisonous alike, and Thom collected the bodies daily in a bucket when the executioner was finished.

He was also a marvelous chaperone for teen-age parties.

Craig was, by then, fifteen and when he had his first party, Van and I were prepared to be alert and much in evidence. We were hardly needed. Ichabod had the situation well in hand.

Since the party was on the terrace out of doors, it disturbed the crane's slumber so he joined the festivities. The first couple who wandered away from the crowd into the bushes emerged in full flight. Ichabod was right behind them with his wings flapping and his bill clacking like a teletype machine. There was no further trouble.

As months passed, his clipped wing feathers grew back but no one noticed. We were entranced with the feathers that had turned pink on his head and breast for the mating season. We would, I'm sure, have hunted a mate for him if we'd known where to find one. It was impractical anyway, because we didn't really know whether Ichabod was a he or a she. I suspected that he was a female because she loved Van, Thom and Craig and hated Molli, Tong Sook and me. Lee and Christy, I guess, were too young in his-her mind. Ichabod tolerated them. Van preferred a "he," so we generally regarded the crane as masculine.

At any rate, he or she got tired of waiting for us to provide and flew over the garden wall. Thom saw the departure, sounded the alarm and the family took off in pursuit.

Our fence was always ringed with Thai faces peering over to view the crazy Americans and to see brindle Pruno, who I think they thought was a tiger. As we streamed through the gate, Thom and Craig were in the lead, with Lee, Christy, the cook, the coolie, the wash ayah, Pruno, Yorick and I following. The bodiless faces around the fence became people and they joined the chase like the tail on a kite.

We raced by two houses of ill-repute without collecting more followers, but in front of the opium den, three men in various stages of dishabille joined me at the end of the line, grinning happily.

"Speak Engrish," said the one closest to me, proudly poking a finger into his chest. "You talk, I risten."

"I can't talk," I panted. "I'm too tired." I wasn't going to encourage a relationship with this high-flying segment of the Thai population.

"Where we go?" my escort demanded.

"I pointed at Ichabod struggling to stay aloft ahead of us. His long unused wings were failing, thank heaven.

"Ah so." My recent acquaintance was either sobering up, drifting further away or losing interest. He stopped running. The tail of the kite faltered and, as I looked back, became an untidy heap of flailing arms and legs. "Oh, dear," I thought, "I hope no one is hurt," but as the followers began to sort themselves out and to again join in the chase, I ran on, leaving them behind me.

We left the lane, leaped over a klong, and ran through a

field or two until, at last, I saw Ichabod come to rest in a tree inside the walls of a crematorium

The Thai crematoriums are not like ours in the United States. The bodies stay crushed into jars until, depending on the class of the deceased, the time comes for the actual incineration. The Queen Mother was not cremated for two years. Our cook's mother was done in two weeks. In the meantime, the relatives have a series of parties and functions to honor the dead while the corpse stays in a jar in the odoriferous crematorium.

The burning itself is not fragrant, and a cremation was in progress as we burst through the gates. The participants and the monks in their saffron colored robes, viewed us at first with alarm and then with interest. It became apparent that we were more fascinating than the ceremony because the crowd left what remained of the remains smoldering atop the funeral pyre, and joined us under Ichabod's tree.

After much bowing and explanations, Craig and Thom climbed the tree to rescue the exhausted and willing Ichabod.

The mourners, I think, invited us to stay for the rest of their ceremony, but we thanked them, bowed nearly to the ground, and left. Still bowing up and down, we backed through the gates into the three opium den denizens who had, I presumed, arrived to give us unwelcome support. They crowded through the gates and decided to stay for the drinks sure to follow the cremation.

Thom, with Ichabod tucked under his arm, swung around, and the human sea of followers parted to let us though. On the dusty, rutted road, the undulating motion, as expected, upset the bird's tummy and he scattered his undigested fish with the abandon of a flower girl scattering rose petals. He never tried to fly away again.

When it was time for our family to leave Bangkok, there was no question of bringing Ichabod home with us. He couldn't have survived the cold winters.

While we were still debating what was best for him, a crane of the opposite sex (whatever that was) flew over the compound, fell in love with our handsome bird and joined him in his tin tub. Since now there were two, and they probably would have a family, we felt it would be safe to take the pair and release them near a klong far out in the country.

We let Ichabod's wing feathers grow and a few days before we ourselves flew away, we drove the birds to their freedom. When we got out of the car and carried them toward a narrow stream, Ichabod, as always, gave up his lunch in our honor.

When we reached the water we deposited the lovers on the bank; they stepped in and, without a backward glance, went squishing away wing to wing.

XIV

The two years in Bangkok went by in a kaleidoscope of heat, parties, friendships, almost unbearable pity for mistreated animals, and, above all, fascinating participation in a way of life so different from our own. The sights, sounds, smells and feelings allowed us to experience a culture sometimes wise beyond its ancient age, and sometimes, as is the way with all cultures, sadly ignorant.

When the time neared for us to come home, I won't say we weren't overjoyed, but if the logistics of the trip over were difficult, the plans for going home were monumental.

Penny and Pogo were not to be left behind, of course. I'd spoiled them for jungle living, they'd never been attached to a wire, and they thought they were members of the family. What else could be done but take them home?

"Well," my husband said in resignation as we plotted our departure, "we'll take them home, but not on the ship. We're going to have an animal-free voyage."

"That will require some logistics." My mind went into gear. "The dogs and Tucky can be shipped by air to Dr. Burns (our vet in Virginia); they'll be happy there for a month, but Penny and Pogo . . . well, I'll work on it." I did.

The final plans went like this!

Ampon had, by that time, become indispensable, so she was eager to come home with us. She had a six-year-old boy, Dang, so of course, he would come, too. If, after she had been in the USA for a month or so, she thought her husband, Sampow, would like it, he would follow. Now, we'd picked up three more humans in addition to our two very small partial-humans. The Walnut Lane house in Vienna, Virginia, large as it was, was too small. We set about selling our old house and building another that would accommodate us all. Naturally, this was all done by letter.

We had a dear friend in real estate business in Vienna who was also a builder. He had eighteen acres on Meadowlark Road, upon which he would build us the house of our dreams, and it would be ready when we arrived in five months. Not only that, he would also sell the Walnut Lane house for us. What more could we ask? We decided to do it.

Letters flew back and forth by nearly every post, but we never had a thought that the house would be less than perfect. Knowing Mac, it would be more than we ever asked, and it was. The room for the gibbons became famous in Vienna long before anyone knew what a gibbon was, and the builders were driven crazy by cars driving up with occupants demanding to see the "baboon room." Maybe the people were just sight-seeing, or perhaps they wanted to be sure that these creatures, that might be as large as Gargantua and twice as fierce, were to be safely incarcerated.

Now we had nothing to worry about but the travel plans. Dr. Burns, the vet, would be happy to have his furry friends for as long as it took us to get home, and my dear friend, Dorothy, would pick them up at the airport and deliver them to the vet.

The gibbons were different. Ampon solved that. She

had a little house in Bangkok and she would take Penny and Pogo there with her. As soon as we cabled that we were home, she would put them in the box we had made especially, and ship them off by American Airlines. She and Dang would then follow.

It all sounds simple as I write, but the shots and the visas and the passports and the reservations and tickets and the miles of paper work consumed the entire final five, hot months we were in Bangkok.

During all this time, a worry lay heavily in my heart. Pruno was twelve, a pretty advanced age for a Dane. Were we being selfish even to plan to put him in a crate and send him off in the belly of a noisy plane? He would be separated from us for a month, and even though he knew Dr. Burns, this might be traumatic.

He grew more feeble every day and one night Van and I made our decision. We felt he couldn't make the trip home as much as we wanted him to. He might die on the way, lonely and frightened, or soon after his arrival. We loved him too much for that, so we made plans to ship Yorick and Tucky, and gave Pruno as much happiness as we could in his last months.

We never consciously picked a date but it was fixed for us. On the morning of our eighteenth wedding anniversary, July 27, 1958, our beloved dog's hind legs gave out and he couldn't get up from the living room floor. It was time.

One of Van's captains had been a medic in WWII, and he came at once. Dr. Burns had given us anesthetic and a syringe to take to Bangkok with us in case there was a necessity for it, so we were physically ready for the injection that would release our kind and gentle family member from his limitations.

I sat on the floor with Pruno's great head in my lap and

Van sat beside me, both of us stroking his warm brown head and body, and our silent tears flowed and dripped onto our hands. His brown eyes looked into ours with such love and trust, and as the anesthetic flowed in, our precious companion floated peacefully away.

I had thought to leave nothing of myself in that land where cruelty and callousness toward animals was a way of life, but a piece of my heart remained behind buried with Pruno beneath the mimosa trees.

The day we buried him I felt the ache of his loss so strongly, but I knew that someday the memories of his joyous life with us would fill the vacant spot in my heart, and he would gallop beside me, strong and free, all of my days. And so it has been.

XV

The sky in Virginia was a brilliant blue, the air was so clear we seemed suspended in it like figures in a glass bubble. As the new station wagon followed the curving road and passed a tiny stream loaded with watercress, we all looked to the left for the first glimpse of our new house.

There it was on the top of a hill, long and low and white, its bright blue roof and blue shutters blending with the sky.

We stopped at the end of our driveway to absorb the wonder of the afternoon and to wrap ourselves even more closely in the feeling of euphoria. We were home! This was our country and before the five of us, was the one, small piece of the United States of America that, for however long we chose, would belong to us.

Huge trees dressed in autumn's gold, red and sunny yellow covered the hillside and enveloped and protected the house. A flowing river of green lawn wound its way up from the road and led our eyes and our hearts to the front door.

There, Hugh McDiarmid, our friend and builder (better known as Mac), and his wife, Dorothy, stood to meet us. Wrapped in our warm sweaters, we joyously took our first tour of the perfection that was our home.

The house we built by letter as we first saw it.

From the beautiful, carved fleur-de-lis that Mac had ordered for each panel of the double front door, to the simple gibbon room with its raised sleeping cupboard on the wall, there was no fault. With borrowed blankets, we slept on the soft carpeting of our own rooms that first night.

Starting early in the morning, the furniture arrived from storage. Even more important, Van and Craig went to Dr. Burns's and brought Tucky and Yorick home. They also brought, from our old house, our other cat, Putty.

Putty hasn't entered the tale before because we had erroneously thought he would be happier staying in the house in which he had been born and so had left him behind to live with the renters. When we left, the renters told us, he went into the barn and there he lived and was fed for the entire two years.

What a joyous reunion! It wasn't necessary to keep the cats shut in until they adjusted to a new place. The doors were wide open all day while the moving men came and went. Both cats, as soon as their favorite chairs were brought in, curled up and slept as if they had been awake forever.

Yorick's plumy tail, his best feature except for his luxuriant, long lashes, waved in perpetual motion. He was so torn between which family member he would follow that we feared he might become neurotic. Finally, Craig and the girls went up the road to see some of their old friends and Van tramped off to explore our eighteen acres of woods. Yorick couldn't resist Van's invitation. He was home in a land he understood, he was free of a cage, and, in the fresh, clear air, he was cool. He went with Van. I watched from the kitchen window until they disappeared, Yorick's shiny black nose snuffling and exploring in the fallen leaves.

By nightfall, aside from the mountains of unopened boxes stacked in the garage and the shipment that was on its way from Bangkok, we were settled; and, exhausted, could go to sleep in our own beds.

The final phase of our homecoming plan went into effect the next day. We cabled Ampon. She shipped the gibbons from Bangkok by American Airlines and, two days later, while we were at National Airport welcoming Penny and Pogo, she and Dang flew to Hong Kong to catch the ship for America.

Our family had grown in two years, but was not complete. Longing for Pruno and wishing he could gambol in the cool air and explore the woods with Van and Yorick, I wondered if we could ever be truly complete again. A month and a half had passed since Pruno's death, and the pain had dulled to an ache as I remembered the mound of dirt under the trees in a country both Pruno and I had found so cruel, but it would do no good to mourn.

To finally let our big dog go, I went out one day into the woods and there, sitting on a log at the foot of a two-hundred-year-old tulip poplar, I opened my heart and felt him bound away. He was all right now, galloping somewhere in joy and comfort, and I was all right, too. I knew that "somewhere" was never too far for my love to reach.

Now we had to get ready for Penny and Pogo. The new children's shop in Vienna was fascinated as I bought infant undershirts, new born flannel sleepers, sweaters and caps. These weren't for indoor wear. Unaccustomed to cool weather, the gibbons had to be warm when they played outside. The store owners and the other customers who crowded around with suggestions as I chose the clothes, would have outfitted them as beautifully as they did their own children, but these were live apes and not toys, so their clothing was utilitarian and nothing more.

The blankets were folded in their sleeping compartment which was a cupboard hung on the wall of the gibbon room. The sun lamp was installed in their ceiling. A jungle gym promised to give the gibbons plenty of exercise on rainy days, and a drain was in the center of the tile floor for easy hosing. The water faucet and hose, however, were outside the room, and enclosed, so the gibbons couldn't turn it on. If they could reach something—they tried it. If it made a mess—they loved it. We were ready—we thought!

Our furry children required nothing in the way of special food. A supply of cucumbers and rice was waiting, and the manager of the super market promised to save, and sell us at a discount, the old bananas too ripe for anyone else. Over-ripe bananas have more flavor, and I knew the gibbons, used to the world's tastiest fruit, would spurn anything picked and shipped green.

When gibbon-arrival day came, Craig, Lee and Christy

were already in school. Van and I made the trip to the airport alone.

The plane touched down just as we drove up in front of the Air Freight building, and we were standing at the open door with an American Airline employee when the familiar cage was unloaded.

A friend of Ampon's had built the crate for us to our specifications and it was sturdy and heavy. It had feeding and water bowls attached to the wall on the inside, with a slot above them so food and water could be inserted without opening the door. We thought, with Pogo's abilities and ingenuity, the gibbons would be safer with NO way to escape. The cage was divided into two parts, a sleeping compartment in the rear, where they could hide and feel safe, and the feeding, exercise and bathroom section in the front. The face of the box was of heavy wire mesh and the remainder was solid.

As two men carried it toward us by the handles on either side, there was no sign of an animal, but when the crate was placed on the floor beside us and I quietly called their names, I heard their joyous squeaks, and the gibbons burst from hiding.

"You aren't going to let them out here, are you?" one of the women employees asked, as I started to open the cage door.

"Oh, they'll be fine," I answered as I reached for the leashes that were attached to one of the handles and the door swung wide. As expected, the gibbons were clinging around my neck in seconds, alternately making cooing noises and grinning. Penny never loosened her grip even when Van spoke to her, although she did turn to look at him, but Pogo, the imp, had been confined far longer than he felt was proper and when freedom was offered, he took it.

Before I could snap the leash to his belt, he leaped from my neck to that of the pretty young lady who had originally questioned my opening the door, and in less than a minute, proved she had been right to be apprehensive.

Of course she screamed and began to push at the small furry creature attached so lovingly to her neck. The three, brave, male employees backed away.

"Close the outside door!" Van yelled. "Pogo, come here!" Penny, Van and I started in pursuit of the acrobat who flew from the soft neck to a pile of boxes. Naturally, the smaller parcels on top scattered. The employees raced to collect them but kept their eyes on the energetic gibbon. Pogo leaped to a counter and ran to the opposite end where a typewriter interested him just long enough for him to get all the striking rods standing upright and entangled. It took but the blink of an eye, with, in essence, four hands operating.

Behind the counter, a honeycomb of large, lettered open compartments, loaded with items waiting to be called for, rose halfway to the high ceiling of the metal, barn-like building. Pogo went up as smoothly as if he were on a fast moving escalator.

By then everyone in the office was yelling or running and, in some cases, doing both. The door opened and two more men bounded in, demanding to know what was going on. No one answered, but they figured it out and closed the door to prevent Pogo's escape. They then joined the melee. Penny, still clinging to my neck, added her touch to the uproar. She began to whoop loudly at everything, and the sound, so close to my left ear, caused excruciating pain. I, too, screamed. Penny whooped louder.

Pogo hesitated a fraction of a second then he, too, became vocal and between the whoops and the yells and the screams and the clatter of objects being tossed from the honeycomb, we had full-fledged, self-feeding pandemonium.

Pogo had been running amok in his honeycomb and, at the same time, watching the antics going on below him with extreme interest. I know for a fact that never in his life with us had he been afforded such a spectacle; not many have. Exhausted, he finally stopped moving and sat down to watch.

The horde below, also exhausted, became silent and stood looking up at the less than twelve-inch-tall ape. Pogo grinned. The multitude sighed and sheepishly assessed their danger from the pretty, fluffy, cream colored creature sitting so benignly above them. When Penny spoke softly and Pogo came down, I prayed this unusually kindly humor would continue.

While we apologized and helped pick up the packages and people's possessions to pile them on the counter, Pogo, with his leash now firmly attached, made his rounds of the employees.

The tiny master magician enchanted his public and they were reluctant to say good-bye. Only when we invited them all to visit in Vienna at any time and to bring their families, did they let us go. They had a tale to take home that night and so did we.

"Well, the gibbons seem to be in good shape," Van said as we drove along the George Washington Parkway on the way home. A car started to pass us and then slowed to our pace. Van turned his head to look at it. "What are those people staring at?"

I looked at the car traveling parallel to us. A man,

woman and little boy were smiling and pointing (two expressions I grew to know well in the following years as Pogo and I, by invitation, made the rounds of schools and clubs together).

They must have been watching Pogo since Penny still clung around my neck like a loving limpet, almost out of sight. I turned around to see where, in the back of the station wagon, Pogo was performing.

He was running back and forth on the back of the seat behind us, alternately viewing the Potomac River on our right and flirting with the people in the car on our left. In this fashion we rolled along toward Route 123 until cars piled up behind us and began to honk. Pogo ran to the back window to see where the noise originated; the car beside us reluctantly moved ahead and took to leading the parade.

A long, green convertible with the top down, took its place. Four young people in their late teens or early twenties rode beside us, laughing and waving at Pogo who was putting on a wonderful performance. He had found a polishing rag and was playing his towel game, staggering around with the rag over his head. When the cars trailing us tired of being blocked in and the honking began again, cars once more changed places. The convertible was replaced by a long, black Cadillac. Two elderly ladies, riding behind their chauffeur, rolled down their window and, without taking their eyes from Pogo, engaged Van in conversation.

"What is that little animal," the elder of the two asked?

"It's a gibbon," Van answered without removing his eyes from the road.

"I hope I'm not being impertinent," the lady said and leaned forward so her head in a flowered hat was nearly out the window, "but what is a gibbon?"

"It's the smallest member of the ape family," Van replied. "These are babies."

There was a quiet conversation between the ladies who appeared to be sisters. The face in the flowered hat reappeared in the window. "We do hesitate to ask, and you must answer us quite truthfully, but would it be possible for us to go someplace to view this enchanting little creature at closer range? You see, my sister and I love animals and we so rarely. . . . Oh, dear! What must you think of us? Do forgive me!"

"There is nothing to forgive," Van said and smiled. "My wife is writing our address and phone number and do feel free to come whenever you wish."

I found a scrap of paper in my purse and a pencil stub in the glove compartment. Penny's nimble hand snatching at both made writing legibly almost impossible. I handed the results to Van. He extended his arm out the window.

"We'll never see them in Vienna," I said. "They won't be able to read my writing."

A gloved hand came from the Cadillac to receive the directions. "Drive closer, Edward," I heard the lady say to the chauffeur. "I can't reach the paper." Edward obliged and Van stretched his arm to the fullest. The exchange was made.

"Oh, thank you so much!" the soft voice said. "You may be sure we will call." The horns started to honk again. In that fashion we went up the Parkway, accompanied by honking and secured by an ever-present, but always changing, automobile on our left until we turned off on the two-lane road that, before Dolly Madison Highway was constructed, was Rt. 123.

As long as Pogo had an audience, he performed. When we turned off and the cars disappeared beside and behind us, he came up to the front of the car, flung himself on my lap and had a nap.

It was one o'clock in the afternoon when we got home, but in Bangkok it was five o'clock in the evening of the next day. As soon as the gibbons had something to eat and had investigated the lowest level of the house to which, when they were indoors, they would always be confined, Penny stretched herself across my lap as I sat on the sofa in the large family room. The room's warm beams and huge seven foot fireplace lulled us both. Just as we had in Bangkok, we began the pre-sleep cuddle period.

Pogo, never willing to give up, tried to continue his investigation of his new surroundings, but he was too sleepy and when Penny got down from my lap and took his spindly arm to tow him to their room with the sleeping cupboard, he offered only minor resistance.

In Thailand, there is no twilight. It's light—and then it's dark, so gibbons in the wild, to avoid predators, go to bed promptly at five o'clock. Penny and Pogo had always had their supper at four and then Penny lay on my lap while I stroked her head and back. Every day at five, on the dot, she had slid down, taken her banana and we had ascended the stairs together. After Pogo came, the transition hadn't gone quite as smoothly. Always reluctant to give up his playtime and not having lived long enough in the wild to develop a routine, Pogo offered resistance when Penny tried to haul him up to bed. In the end, she always won just as she had now succeeded in putting him to bed at one in the afternoon, but, like most things with Pogo, sleepy or not, it was a tussle.

Van and I followed the gibbons and were gratified that they had accepted the sleeping cupboard as their safe haven. They might just as easily have chosen the shelf in

Penny and Pogo in their sleeping cupboard.

Ampon's closet which was next to their room, but we'd done something right and they were content. Penny arranged their blankets and they were asleep in seconds.

Their room opened into the laundry area and I closed their screen door and locked it from the outside.

Tomorrow would be another day and though this one was ending early for the gibbons, having traveled with small children from time zone to time zone, I knew that Penny and Pogo would gradually adjust and in a few days would go to bed at the "real" five o'clock.

All I really had to do I thought, was to acclimate Penny starting tomorrow, and she would see that Pogo conformed. I had no idea what a horrendous day "tomorrow" could be.

XVI

It was Saturday morning. The clock on my night table said it was six-thirty, barely dawn, and obviously something was terribly wrong with our furnace. Click . . . roar! It turned on. Click . . . whoosh! It turned off. Click . . . roar, it was on again. Click . . . whoosh, it was off. As in the night before Christmas, "I sprang from my bed to see what was the matter."

While I was searching for my robe in the faint light, Van sat bolt upright. "What's wrong with the furnace?" he demanded as he leapt out of bed and rushed to close the window. He didn't expect an answer. The oil-burner continued its cadence.

We both raced down the stairs, struggling with the arms of our robes and tripping over the belts and each other. In the kitchen, directly above the furnace, the noise was, of course, louder. Van bolted through the door to the garage and then on down the back steps to the lower level. I was a step behind him. He should have come out right beside the furnace, but the door was locked. As in a Mack Sennett comedy we went through sudden stops and pile-ups, bumps and crashes, until, disentangled, we flew back up the stairs.

All the time I was worrying about Penny and Pogo, locked in their room and probably cold. From the warmth of the two main floors of the house, it was obvious that it was the lower level that was in trouble.

I needn't have worried. When Van unlocked the door from the hall to the down stairway, Penny (supposedly incarcerated in her room) climbed me like a palm tree, making her happy squeaking noises and smiling broadly. The tip of her pink tongue stuck out of the corner of her mouth between two even row of white baby teeth and she looked so engaging I forgot that, since the gibbons were loose, there was bound to be damage. I hugged her to me and followed Van down the stairs.

At the bottom, Pogo hung from the thermostat on the wall. Its cover lay on the floor and Pogo's little, black finger was flicking the mercury switch time after time after time. As the furnace came on and went off with thumps and hisses, Pogo's pride and enjoyment knew no bounds. He grinned at me in innocence, but in the shining, shoe button eye, I saw that gleam of deviltry. He knew exactly what havoc he was wreaking.

I plucked him from the thermostat and he came willingly. After all, he'd made his point and, not only that, he'd gotten us up and here we were to fix breakfast and to entertain him.

Van determined that the thermostat was now totally inoperable and as he taped the cover in place for its protection only, I started out to see what other damages we were going to have repaired on a Saturday, at time-and-a-half pay. I could hear Van muttering, "Damn monkeys," and I didn't blame him one bit.

The family room wasn't in bad shape. I'd made the floor lamps out of pot-bellied stoves of cast iron, with painted milk pails for shades. They'd withstood the onslaught. The hunting prints were bolted to the wood-paneled walls and were intact. Even the draperies, carefully gibbon proofed, survived, but the laundry area, Ampon's room and the furnace room were a shambles.

Since the gibbons had never damaged the window screens in Bangkok, we'd felt that a normal screen door on their room would let us keep better track of them when they were locked in, and would imprison them when their presence was not required.

The solution was totally inadequate in Vienna, Virginia. The screening had been driven outward by gibbons launching themselves from the jungle gym and hitting the screens with both feet. We learned this was what had happened, after we covered the door with hardware cloth and listened to the bangs and thumps for the hours it took the gibbons to understand that this time their ploy would not work.

Ampon and Dang's room, attractively made ready for their arrival, no longer had curtains at the window. They lay in a heap on the floor. The bedspreads made another pile, and the carpet and beds were decorated with all the knick-knacks gathered not only from that room, but from the family room, the laundry room and the contents of Van's extensive tool box. A squashed banana topped the mess.

There was more damage in the laundry section of the basement. A clothes-chute from the two floors above emptied into a basket next to the washing machine. Soiled clothing, table linens, sheets, socks and what ever else had been in the basket, hung over the furnace and littered the floor. The sight would have discouraged Pollyanna. "Coffee," Van mumbled. I agreed that we could plot, plan and cope when we were calmer.

I attached Penny and Pogo's belts and leashes and, for one of the few times in their lives with us, they got to go upstairs. By then it was seven-thirty and no longer too early to call the plumbing and heating contractor.

"Mr. Utterback," Van said, after his first swallow of coffee, and a cigarette, "we have a problem."

"Oh-ho!" Van reported that the plumber wasn't surprised. "I expected we might. What is it?"

"The gibbons have broken the thermostat and it's cold on their level," I heard my husband say. "We need help! When can you come?"

Apparently Mr. Utterback was ready for us and, indeed, he told us later that he and Mac McDiarmid had anticipated there would be difficulty when the apes arrived, and had saved Saturday and also Monday for us.

"He'll be here within the hour." Van turned momentarily from the phone.

"Bless him," I said and tried to keep Pogo's grasping hands from the phone.

"By the way, Mr. Utterback suggests we'd better get the carpenters too. We're going to have to enclose the thermostat or this will only happen again I suspect." There was a pause after he talked again with Mr. Utterback. "He wants to know if the animals will be loose when they get here," Van said to me. "It's all right, Mr. Utterback," he told the plumber. "They're not vicious. Oh, yes. They're safe for children. They love children. Okay, we'll see you in an hour."

The conversation with the carpenter, Mr. Jackson, went much the same, and I had the feeling, that the workmen were all eager to get out on Meadowlark Road to be the first to see the gibbons. After all, the local residents had driven them crazy during the building of the house wanting to see the "baboon" room. Now they'd be in on, literally, the ground floor.

The phone rang before I could get the gibbons' breakfast ready. Van had sat down with his coffee. I answered it.

"Good morning, Mrs. Vanderhoef, this is Mr. Blosser." Mr. Blosser was the painter.

"Good morning," I said. I wondered what excuse he would find to join the rest of the house-building crew. "What can I do for you?"

"I thought I might run out to make sure everything is to your liking." There was a pause. "That is, if it's all right with you."

"Of course it is, Mr. Blosser. How thoughtful of you to be concerned." I waited while the nice man "ohed" and "ahed" a second. "By the way," I added, "perhaps you'd like to bring your kids to see the gibbons."

"Oh, that would just be wonderful," Mr. Blosser said, "but they'd be my grandkids. They've been driving me crazy ever since they heard the monkeys were coming."

I would certainly have to educate this town. These were apes, not monkeys.

"That's fine, Mr. Blosser. How old are the children?"

"My son has two little girls. They're seven and nine."

"That's perfect," I said. "We'll see you in a while."

Before I could walk across the room the phone rang again. "News certainly spreads," I remarked.

It was Mr. Utterback's helper, Gregory. "Mrs. Vanderhoef, can I bring my kid with me?"

"Certainly, Gregory!" I answered.

"Thanks! See ya," he said and hung up. That was my preferred type of conversation.

Lee appeared in the kitchen. "What's wrong with the

telephone?" she asked "What crazy people call anybody at seven-thirty on Saturday morning? Was it for me?"

I shook my head. The phone rang again.

"Hello," I croaked as I unwound one of the leashes from around my neck.

All this time Penny and Pogo had been pulling on their tethers, bounding onto and running along the counter top to turn on the faucet or peer out the window above the sink. I had found myself bound by either the leashes or the phone cord as I twisted and turned to keep control of the two prototypes of perpetual motion.

"Hello." The voice on the other end sounded tentative. "Is this the Vanderhoef residence?"

"Yes, it is," Pogo swung by on the cupboard handle behind me and grabbed a handful of hair. My whiplash that had almost healed in the month and a half of rest, was back. "Ouch!"

"I hope this isn't a bad time to call." The voice on the other end of the wire was apologetic. "I hope I didn't waken you."

I laughed.

"This is Dave Thompson," the voice continued. "My wife and son and I drove beside you on the George Washington Parkway yesterday."

"Oh, yes." How did they know who we were? I asked the question.

"We traced your license plate," Mr. Thompson answered. "I'm an Arlington policeman. I know I would hate it if anyone did this to me, but my son won't let me alone. I wonder . . ." there was an embarrassed pause . . . "I wonder if there is any way we could get to see that little monkey again? By the way . . . what kind of a monkey is it?"

I tried to keep irritation out of my voice. "Pogo is not a

monkey, he's an ape. He's a gibbon and, yes, if you would like to, you may bring your son out to Vienna to play with Pogo. Actually, we have two baby gibbons, but Penny was hanging onto me yesterday so you probably didn't see her. When would you like to come?"

"How about this morning? It's a nice day."

"Fine," I said and thought about the carpenters, the plumbers, and Mr. Blosser and his grandkids. Oh, well it couldn't get much worse. We might as well get it over with in one day. (I didn't then know it would go on like this until the snows came!)

Craig and Christy staggered into the kitchen. Craig was dressed in jeans and a sweat shirt, but Christy stood shivering in shortie pajamas, no robe and no slippers.

"Get dressed, Christy," I said automatically before I even said good morning. "A lot of people are coming out here and you'd better be ready." She went off so docilely I wondered if she might be sick.

Mr. Utterback came first. With him was his wife, her sister, his brother-in law and two neighbors. Still in my bathrobe, I hurried outside into the, fortunately, over seventy degree, sunny morning and unsnapped the leashes from the gibbons' belts. We weren't about to entertain in the house with coffee. There promised to be a legion assembled by nine o'clock, and I wanted the visitors to be kept outside at all costs.

The trees in the yard were mostly tulip poplar, hickory and oak, well over a hundred years old and all straight and very, very tall. Penny and Pogo were up and swinging in the branches before I could lay the leashes aside. Van went in to get dressed while the Utterback group and I craned our necks to watch the gibbons above us. We had no trouble hearing them. The exuberant whoops echoed endlessly.

Our nearest neighbors were at least five acres away, but the sound carried. A Volkswagen raced up the drive, came to a stop and those neighbors, fortunately good friends, Betsy and Craton Guthrie, tumbled out.

"Where are they?" The airline pilot and his wife stood looking up into the trees. They needn't have bothered.

Penny and Pogo loved to greet cars. Down they came and before I could shout a warning, had snapped the radio aerials from the Volkswagen and from the two trucks that had brought the Utterback tribe.

With nothing left on the vehicles on which to swing, Penny took stock of the strangers. The minute her shining, black eyes lit on Craton Guthrie, she was in love. She twined her arms around his neck and showered him with coos and kisses. Craton cooed back. Pogo had no idea what was occurring, but because Penny was obviously smitten with Craton, Pogo wasn't to be outdone. He draped himself on Craton's head, clasped his arms around the poor man's forehead and eyes and lay, like a fur hat, grinning at the assembled audience, who loved it!

A car drove slowly up the long driveway and came to a stop in the parking area. The gibbons started for the car at a dead run.

"Put your radio aerial down!" Van yelled at the driver.

Too late! Snap, clatter. The long, shiny rod fell to the ground.

The Thompson family sat in the car, mesmerized, staring at the two little apes running around on the Plymouth's hood. Dave Thompson opened his window and (as in the old saw "I opened my window and in flew Enza") in flew Penny and Pogo. The car emptied quickly.

It was a repetitive morning. Cars drove up and were de-aerialized; then the occupants stayed to watch as Penny and Pogo swung freely from tree to tree. Sometimes the visitors wanted to have a chance to hold one or the other of the elusive gibbons.

Craton, as smitten with Penny as she was with him, took over the outside watch while I dressed. I then joined Van, Mr. Utterback and the rest of the carpenters and electricians who arrived with their families, friends and relatives, to discuss the modifications necessary to gibbon-proof the lower level.

The work was to begin as soon as the materials were collected. Everything—laundry, furnace and thermostat—would be enclosed behind louvered, bi-fold doors which would have stout latches like the one we'd had on the fruit cupboard in Bangkok. The door to the gibbon room would be faced with hardware cloth glued, nailed, stapled and bolted to the frame. In other words—nothing would ever remove it. We weren't sanguine enough to feel entirely secure in our decisions, but they were the best we were prepared to make. During the consultations, no fewer than a dozen people, mostly children, wandered in looking for a bathroom. Ampon's was handy, so that's the one they used.

The workmen, after a great shifting of cars in the parking area and up the driveway, left to gather the lumber and other requirements, and to take their families home. Theirs was the only exodus. As they drove out, other vehicles turned in. The word was out that the "baboons" had arrived and all of Vienna wanted to see what they looked like.

Late in the morning, a black, chauffeur-driven car came up the drive. The two ladies of the Parkway peered eagerly out of its open windows. "We tried to call you," said the

Penny adores Craton Guthrie and is not shy about showing it.

one whose name we learned was Marcia Christianson, as Penny and Pogo leaped on the shiny hood, swung around the aerial and snapped it off, "but the line was always busy or else no one answered. We decided to take a chance. Oh, dear, Adele," she turned to her sister. "We seem to have interrupted a celebration or a gathering of some sort." They both looked at the throng lining the driveway to watch the gibbons dismantle another antenna.

"Oh, it's a gathering, all right." Van laughed and assisted the ladies from their chariot. "Everyone is curious about the gibbons. Come join the mob." He escorted them up the steps from the parking area to the lawn above.

Penny and Pogo recognized genuine feeling in the two elderly ladies and each gibbon reached up a hand to the lady of choice and was swung up to nestle in a soft, crepey neck that smelled, I'm sure, of lilac water.

It has been one of my greatest regrets that no camera preserved Marcia and Adele's expressions as the furry arms reached around their necks and the small heads lay on their shoulders. With no thought for their elegant clothing, they each hugged a little body to them in a moment of pure, unadulterated glory. The crowd watched in awed silence.

Mr. Waters, the mail carrier, turned in, and seeing the multitude on the lawn, drove slowly up the drive. Penny and Pogo pushed themselves from the loving arms and, hand over hand, swung from one small dogwood tree, under the great canopy of poplars and hickories, to another. They snapped Mr. Waters' aerial and dropped it onto the gravel half-way up the lane. Sensing company for their own misery, the crowd cheered.

The car windows were open and in the apes went. In a heart beat they were back out the windows and up into the trees, each clutching an arm load of letters.

Mr. Waters leapt from the car in pursuit of the mail entrusted to him, but no one noticed that he still had not closed the windows. While everyone, the ladies included, ran in undisciplined frenzy to gather the envelopes drifting from the trees, Penny and Pogo quietly swung down, gathered more letters from the seat of Mr. Waters' car, and the "mail drop" continued almost without interruption.

Tired of the game at last, the gibbons began breaking off dead branches and throwing these down as gibbons do in the wild. This is part of their self-preservation. When pursued by a predator, they can't chance having a rotten branch drop them to the ground, or even slow them, so they do their pruning in their leisure time.

As the letters rained the carpenters returned with their truckload of lumber and louvered doors. Some of the onlookers had to move their cars so the truck could get in. There was much backing and changing positions, but no one left.

When the drifting of the envelopes ceased, Mr. Waters discovered that, aside from newspapers and magazines, his front seat was empty.

"Now I have to sort the letters all over again," he said mournfully. "I hope we have them all."

"I think we will, as soon as those kids bring up the ones from the ravine." I referred to the six sons and daughters of various onlookers who thought Penny and Pogo were staging the paper chase just for them; they were down in the small hollow worn by runoff from the land above. They had been warned not to scuffle in the fallen leaves, so no letter would become buried. Mr. Blosser, the painter, stood on the edge and directed operations, but I had heard giggles and shrieks of glee followed by admonishments, so I thought I'd do some sifting when the crowd thinned. I'd hate to have someone go to jail for non-payment of a bill that was buried in our ravine.

To avoid having Penny and Pogo assist in the sorting, Mr. Waters, his mail, the Christianson ladies, two of somebody's relatives and Van went into the kitchen where the counters would make a good sorting place. Since by then it was almost lunch time, some visitors left, saying they hoped they could come back; some children had tantrums and wanted to stay, but eventually, the yard and the parking area were almost cleared.

Penny and Pogo came out of the trees when I went in to

fix their lunch. I locked them out and they sat on the windowsill peering in the window, and crying. The Christiansons couldn't stand their distress and abandoned the mail-sorting to go outside and cuddle the distraught apes; but as soon as I came out the door, the fickle creatures left the tender hearted ladies and clung around my neck.

Juggling tray and apes and followed by the ladies, I staggered to the chairs under the trees and we all sat down.

By the time the gibbons had finished lunch and had had the cuddle period we always had in Bangkok, the carpenters had finished the gibbon room door.

When Penny got up and took Pogo by the hand, we all, carpenters and the ladies included, trooped behind them to watch them tuck themselves in bed. It was one o'clock in the afternoon and, like Penny and Pogo, I thought it must be bedtime. Tomorrow, I thought. I'll begin changing their time schedule tomorrow.

Mr. Waters finally took his mail and left; but thereafter, though he came daily to see the gibbons, his aerial was always down and his windows were always up.

Marcia and Adele remained friends and came often until Adele died at age eighty-five and Marcia was so lonely she lasted only a few months more. Edward, the chauffeur, went to work for other friends of ours, but no one, he said, could take the place of "his ladies."

The doors and the locks worked, and Van wasn't quite so outdone with the gibbons after the first bill for the repair and restoration came. It said simply, "Monkeys monkeyed with thermostat." Van had the bill framed, and it hangs on the stairway to the lower level amid what we call "The Rogues Gallery"—the framed pictures of our life.

Although the carpenters and plumbers continued to work that momentous day till five o'clock, Penny and Pogo slept quietly behind their, gibbon-impervious door. They must have gotten up to play in the dark after a while, but, exhausted, the Vanderhoefs heard no sound. After all, we had to recuperate for what we expected might be something of a repetition the next day. It was!

XVII

By the time Ampon and Dang arrived, the daily routine had been established. It didn't always work, but for the most part we had a sort of schedule.

Since gibbons are more like small babies than they are like accepted household pets, they can't be left alone in the house while "mother" goes off to the grocery store. That posed a problem.

Only once did I try to take them with me to the supermarket, leashed and diapered, of course. Neither the market nor I are ever likely to forget our brief, few-minutes onslaught.

The leashes were short, but the arms were long, and while I bent to retrieve a dozen or so cans or boxes from the floor, more items cascaded on my back and head. No other customer assisted me. Every customer—man, woman and child—stood, gaping in delight at the creamy-furred demons who were emptying any shelf within reach. Blocking the aisles so that my basket of gibbons and unwanted items could move in no direction, customers giggled and "oohed" jostled and shoved, and all the time kept up a running patter.

"Aren't they adorable?" "Look at their tiny hands and those perfect fingernails." "Gee! Where are their tails?"

"How come they're so clean?" "What do they eat?" A thousand questions flew as fast as the cans and boxes.

A large woman, seeking, no doubt, to be friendly and helpful, opened a box of sugar cookies and gave Penny and Pogo each one. Thus inspired, the on-lookers rustled among their basketed items for something equally as delicious (and as diarrhea-producing) as the cookies. I bolted. The basket was a wedge. In the blink of an eye, it was up to speed and we were out on the sidewalk before our audience knew they'd been abandoned.

Henry, the nice man who helped customers to their cars with the groceries, stood on the sidewalk. "What are them things, Missus?" he asked, looking at the gibbons. I explained, and gathered my exhilarated furry children to me and prepared to go to the car. Henry followed me across the parking lot.

"Do you get'em in the swumps?"

"In the what?"

"The swumps, you know, like down there in Loosianna. Them muddy places."

"Oh, no," I said, catching on. These come from Thailand."

"From where?"

I started to repeat, then thought better of it. "From way across the sea," I said as I got in the station wagon and rolled up the windows. "Bye, Henry. If you'll put the groceries I left at the curb back on the shelves, I'll be in this afternoon to shop, after the gibbons have gone to bed."

"The what?" I heard Henry say as I drove away.

By the time the kids came home from school that day, Penny and Pogo, still adjusting to our time, had gone to bed. Craig drew the job of baby-sitting while I went back to the store for the groceries. That was part of the temporary routine we had established while waiting for Ampon and Dang to sail across the Pacific.

They arrived at last. Ampon, bless her heart, had been seasick from the moment she first set foot on the ship still tied to the dock. Dang, at age seven, said he wasn't worried. He knew, if she died, I'd come get him. They'd both made it, however, and it was such fun to introduce them to the differences in our two countries. The supermarkets, the dishwashers, the electric stoves, washers, dryers, furnaces, vacuum cleaners and TVs were all new and wondrous, unheard of in Thailand at that time.

The colorful fall trees were exciting to them. While we were still in Bangkok, I'd shown them pictures of autumn in New England, but Ampon thought each family had painted the leaves on its own trees. She was excited to know that she'd be able to see the beauty that was Fall year after year.

When the first snow came and she went out to pick up a handful, I watched her expression change to one of perplexity.

"Not warm," she said as she held out her hand with its mound of white crystals. "Cold!"

"Yes," I answered. "Did you think it would be warm?"

"Madame say it white blanket cover ground in winter. Blankets warm."

She was right. It should have been warm.

Penny and Pogo's reaction to the snow was somewhat different. With the help of friends and thrift shops, in the days when tiny humans no longer wore snow suits with legs, we'd found two snow suits that just fit our furry babies. On the morning of the first snow we encased them in undershirts, diapers, flannel pajamas with plastic feet, sweaters, snowsuits, hats and mittens, and out we went.

Ampon likes America.

Pogo rides on Erawan, a teakwood elephant.

The snow was about three inches deep and still falling when the apes took their first steps into it from the covered terrace off the family room. The snow came almost to their knees, and when they were no more than a foot from the dry surface, they stopped. Penny put a mittened hand into the white stuff, shook it violently when she realized it was cold, and ran back to climb up my body and to cling around my neck. "Whoop!" she shrieked in my ear as she glared at the uncomfortably icy world laid out in front of her. "Whoop, whoop, whoop!"

Pogo stood uncertainly, an attitude new to Mr. Confidence. He felt the snow and even bent over to put his face in it; then, with no further thought, he whirled around to join Penny around my neck. He leaped so far he hardly left another foot print in the snow.

It had taken almost a half-hour to get the gibbons and the Vanderhoef household ready to go out to play. Counting the minute and a half outdoors, it took less than five minutes to get everyone "unencoutered."

Usually our established routine worked well. In the summer, when Ampon got up, she turned on the light in the gibbon room and gave each gibbon a banana. They played by themselves until someone was ready to go outside with them. In the trees they were, of course, in their element. They swung, hunted bugs, broke off rotten branches, whooped loudly and exuberantly and behaved as gibbons should. Only if I went back into the house did we have a problem. When that happened, they screamed in terror and came down from the trees to sit on the windowsill and cry until I came out again. I spent as much time as possible in the great outdoors.

In two or three years, we built a stable. Naturally, it filled rapidly. Lee was given a horse; Christy got a pony for

Christmas. (Craig had gone off to college in his Model-A Ford by then.) Two goats, Huntley and Brinkley that we'd acquired to eat the honeysuckle, were joined by two more, Harvey and Harkness. A Sicilian burro named Sebastian came as the result of an ill-advised raffle ticket, but he was accepted and loved.

The in-house pets had multiplied too. Maggie, a white Persian cat, joined Tucky and Putty. Her owners had moved to a "no pet" apartment. A friend found a dog in a field and brought him to us because she thought he was Yorick. We put ads in the paper, but no one claimed him, so Ampon adopted him and we named him Carmichael. He, like the gibbons, lived below stairs except during the day when he preferred the sunny yellow carpeting in the living room. The fact that he was black and shedding did the carpeting no good.

Tucky left us in the third year and, as with Pruno, there was a gaping hole in our lives. He was the first in our little cemetery on the edge of the woods.

Craig, determined there would never be a vacancy in the ranks, drove home from college bringing his cousin, Barb, two friends, and a young cat named Ole Olson. Ampon, being Thai, transposed has r's and l's, so Ole became Orrie. He lived with us for two years until, in the night, he ran in front of the rare car on Meadowlark Road. He was number two in our pet cemetery.

Eleven back acres of the woods were fenced, because the goats preferred the expensive shrubbery around the house to the free honeysuckle. The fence kept the goats where they belonged, but since they wanted to be where they could see us at all times, only the first twenty feet of their enclosure was ever cleared of the heavenly smelling weed. When we walked the paths in the fenced-in woods,

Penny and author.

Penny and the author in one of the ape's rare trips upstairs.

to give the gibbons a change of scene, everything with four legs joined us. Penny and Pogo danced ahead. Huntley, Brinkley, Harvey, Harkness, Sebastian, Folly and Sing (the horses) followed. Maggie and Putty meandered, the dogs ranged and the humans were allowed to walk between the gibbons and the goats, butted gently from behind if they went too slowly. It was a comical parade.

Most days, one or another of my friends came to lunch with me and the gibbons or to walk with us in the woods. Even my friends never quite understood why they came, because they had to protect their sandwiches from onslaught, their iced tea from furry fingers and their necks from whiplash . . . but still they came.

It was essential that I spend many hours a day with Penny and Pogo, to keep them from reverting to wild animals, and, also, they were only content when I was near. They loved Ampon, had some affection for Van and the kids, but I was their "mother" and they weren't old enough yet to "let go."

Finally when we got them on the Bangkok routine of supper at four, cuddle period followed by bed at five, I had a little more freedom. In the winter, when the light began to fade early, apprehensive Penny insisted on supper at three and bed by four.

In the winter, when it was too cold to be outdoors, we spent our time in the family room. There were large beams across the ceiling and I had a series of ropes that I could attach to them with hooks. These gave the gibbons plenty of exercise. Whoever was sitting with them could watch television, if he or she didn't mind having the machine turned on and off constantly, but Penny and Pogo were so inventive and fun to watch that usually the TV ran second best, anyway.

Penny and Pogo at Supper

Calming down.

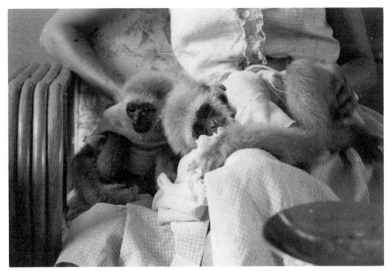

Penny almost asleep—Pogo close—on author's lap.

Pogo, with his rag in his mouth, hangs on a Thai musical instrument.

Often, in the winter, Pogo and I were invited to visit schools. Penny, as time went on, grew unpredictable with little boys and wanted to bite them, so she stayed at home with Ampon, but Pogo loved the trips. He was patient and playful and the small children loved to have him join them in the school yard, on the jungle gyms, swings and slides. We usually went for an hour and ended by staying the whole afternoon. The years flew by.

Dang learned to speak English almost immediately and did well in school. He made friends easily and of course brought home his friends and their childhood diseases. Mumps was first. Dang's case wasn't severe, but Pogo's was. For several nights I sat up in Van's den with Pogo wrapped in a blanket on my lap. For the only time in his life the little gibbon was quiet and wanted to be loved and cuddled. His pansy face was rounder, lumpy and and very, very sore. He was able to eat only rice and soft bananas.

Finally, early one morning as I dozed in the chair, I wakened to find my lap empty. There wasn't a sound, and, in dread, I hurried to see where Pogo was and what havoc he had wrought. I found him tiptoeing around the living room, but not a thing was out of place and the patient felt fine.

Chickenpox came next, and the treatment called for corn starch baths to stop the itching. Pogo thought he was going to be drowned and fought vigorously, but his case wasn't bad and we only had to fight the bathing for a day or so.

By then, when the gibbons were ill, we no longer went through the vet to the pediatrician as we had at first. We went directly to the baby doctor who prescribed as he would for a newborn. (In the end we even gave up the specialist and used our own family physician,

Dr. Newman, who loved the gibbons and spent as much time as he could visiting them.)

Penny got pneumonia and again I sat up in the den. During the day I carried her with me wrapped in a blanket. Thanks to penicillin, the siege was short and she was back to normal in only a few days.

Occasionally we had a day to remember. One of the most memorable was the day we first went (by invitation) to the new Vienna children's store.

As always, when we were out in public, the gibbons wore diapers, and since it was cool they also wore the elegant hand-knitted, cable-stitched sweaters that my aunt had made for them. Penny's was yellow, Pogo's was a deep blue. The gibbons were, of course, on their leashes.

Lee, Christy, the gibbons and I were the only customers in the store, and the first mistake came when the store owner begged to hold Pogo, and Lee handed him over.

Before we could say, "Hang onto him tightly," Pogo gave the woman a quick hug; a quick tug pulled the leash from her too relaxed hand and he was up, up and away. A revolving stand filled with children's sun glasses, barrettes, and other gee-gaws spun so rapidly it whirled off the counter with a horrendous crash and clatter. The store owner, whose name was Marge, didn't know whether to chase Pogo or pick up her scattered wares.

While she stood in an agony of indecision, Pogo flew along a counter that ran from the back to the front of the store. Christy grabbed at the leash as it went past her, but it whipped up and around and she missed it. Pogo leaped from the counter and swung on to a children's clothes tree on which tiny dresses on hangers were attractively displayed. The tree was shaped like a giraffe and we watched as it teetered back and forth. The dresses lifted in the air until Pogo let go and made a leap for the front display window. The clothes tree fell. Naturally it broke. With Penny shrieking in my ear I carefully stepped over the fallen, injured and useless giraffe. After all, we broke it; therefore, we were going to have to pay for and own it.

Penny thought the whole thing was exciting, although she didn't fight to join Pogo but kept up a running "eeeee eeeee" sound that just missed puncturing my ear drum.

Pogo, when he had reached the front window, came to rest on the head of a small boy mannequin dressed in a plaid jacket and a Davy Crocket raccoon hat. There Pogo sat in his blue sweater, observing the passing show on the sidewalk outside, while, at the same time, drawing a crowd.

Just as Lee, Christy and Marge stumbled into the window to capture him, and in spite of my loudly yelled, "Don't open that door," a woman and her little boy did just that.

Pogo was out the door, across the sidewalk and loose in a large parking lot that was filled with a forest of enticing radio aerials. His trailing leash was no more hindrance than it had been inside the store, and we stood helplessly listening to the familiar snap and clatter as antennae, one after the other, broke and fell to the pavement.

We wrote many notes which we tucked under the wiper blades, offering to pay for new aerials, but only three car owners, of the more than thirty, made a request. I can only assume that the others had the fun of joining in the chase and watching the antics as we pursued Pogo through the black-topped jungle.

When at last there were no more cars to decimate, Pogo, who had kept his eye on us most of the time, ran back, and with his arms around my neck, beamed at the gathered crowd. Like the Pied Piper, we led the way into the children's store and the legion crowded in behind us.

The owner told us later her volume of business that day was greater even than at Christmas and she refused to let us pay for the mutilated giraffe. "Bring them back anytime," she said as we wormed our way through the crowd. "I mean that from the bottom of my heart."

XVIII

Penny and Pogo were invited to be the guests of honor at the Tailwaggers Association Fair! The Tailwaggers were, as nearly as I could tell, animal lovers whose association had spread all over the country. They did good works concerning animals, and the gentleman who called let me know how important Penny and Pogo would be to the Tailwaggers' fund raising activity.

"We have heard of your chimpanzees," he had said, "and we could think of nothing more essential to the success of our fair than the appearance of your pets."

At least Penny and Pogo had advanced in stature from baboons to chimpanzees. "What's involved in being the guests of honor?" I asked. "And by the way, these are gibbons, not chimpanzees. They are the smallest of the ape family." My voice must have sounded a little cool.

"Sorry, I didn't mean to insult either you or the little fellows" (why do men always assume everything attractive except a ship is a male?), "and there really isn't much involved in being guests of honor. We have a stage, the emcee will introduce you, you'll tell the crowd something about chim ... er ... gibbons (do I have the right name for them?), answer a few questions and that's all there is to it."

It sounded simple enough except that I would have to

"speak." I have always maintained that no one can beat me talking sitting down, but I'm no good on my feet. I tend to panic when faced with an audience. I was silent for so long as I mulled this over that the man must have thought I'd hung up.

"Mrs. Vanderhoef, are you still there?"

"Yes," I said, "I was just thinking."

"What have you thought? Can we count on you?"

Again I hesitated. I had to force myself to do these things. If people were interested in gibbons, I felt bound to give them the opportunity to learn about them.

"Yes, we'll come," I said feebly.

The directions and instructions, when I got them, were so complicated I began to regret my decision, but I'd made the commitment and now I was bound to fulfill it.

"Ampon," I'd asked when I had put down the receiver, "how would you like to go to a fair in Bethesda, Maryland? Penny and Pogo have been invited, and I could use your help." Privately, I thought she could use a change of scene as well.

"Where dis place?" she'd asked.

I'd gotten out the map to show her, then realized it was the distance she was asking about, not the site. She sometimes got car sick if we went too far.

"Not very far," I'd reassured her. "It will take only an hour or so to get there, then we'll make our speech and then we'll come home." How simple it all sounded.

On the day of the fair, we assembled the necessary equipment. We were glad the weather was bright and sunny, the temperature in the mid-seventies and the humidity low. The only clothing we would have to take for the apes was a goodly supply of diapers. In those days disposable diapers didn't come in preemie size so we had to cut ours down to fit. It made the diapers more flimsy, but for the most part, they did the required job.

In the basket of food for the gibbons we had an ample supply of bananas, cucumbers, rice and a jar full of leaping, live grasshoppers laboriously caught in our garden. Penny and Pogo loved them and we hoped they would choose the protein instead of the sweets the audience was sure to press upon them. Whenever we toured with the apes, no matter how many times I said, "Please do not give them anything sweet," children, whose one idea is to please, always presented the apes with the things they themselves liked... cookies, cake and even chewing gum.

Each time, ever the optimist, I hoped the next time would be different. Still, in the food basket, we had the inevitable Pepto Bismol.

The parking lot at the fair was overflowing with people as we drove up, but I found a parking space not too far from what I took to be the stage. It wouldn't have been too far if I hadn't had my arms full of gibbons and if they hadn't been so intriguing to the milling throng. As we pushed our way through the crowd things began to deteriorate. "Oooos, Aaahs" and "Oh, looks" followed us. Grimy, grasping children's hands tugged and clung to my pastel skirt as children tried to wrest a gibbon from my arms. Penny held my neck in a choke hold. Pogo, ever ready to play, tried, in vain, thank God, to break free. By the time we reached back-stage, I felt, and must have looked, as though I'd been through a violent uprising, and my furry children were keyed to a fever pitch.

I thought we would never fight our way to the stage, but there it was at last, not much more than ten feet in front of us and I felt like a drowning man when he sees a life preserver.

The wait, until the time for our appearance, was interminable. Not only were Penny and Pogo hyper-active, grabbing and grasping for humans and objects, we had to change their diapers twice and I grew more nervous about appearing before that sea of untamed mankind.

"Give them each a piece of banana, please," I said to Ampon. "Maybe eating something will calm them."

Ampon handed out the squishy treat just as an act ended and the emcee announced us. I staggered onto the stage with each gibbons clinging my hair with one hand and mashing banana onto my blouse and neck with the other. We came to a halt as the gibbons joyfully discovered the audience and, unseen by me, dropped their bananas onto the stage floor.

"This is Mrs. Vanderhoef and the two monkeys, Penny and Pogo, that she and her family brought back from Bangkok, Thailand. Mrs. Vanderhoef, would you step up to the microphone?"

It was only the indignation felt at the label "monkey" that propelled my feet forward. One of my feet found the banana. I've seen ice skaters perform, to great applause, the intricate foot work I managed without a moment of training. Coming to a stop near the edge of the stage, near the microphone, I hoped the audience would assume my fandango had a purpose. Surprisingly, it drew a smattering of applause.

As the three of us stood in front of the microphone (one of us viewing it with silence, terror and awe), I could hear the emcee speaking, and it seemed far, far away. We might have remained there, immobile, for the duration of the fair if Pogo, always seizing the initiative, had not leapt from my arms, swung twice around the microphone's pole and wrestled it to the ground.

The microphone, as if in agony, screamed as it rolled back and forth at my feet while Pogo continued to attack it. Penny pushed off from my chest and dropped to the stage whooping loudly in elation. The hundreds of watchers, after a moment of silent amazement, roared with laughter. The emcee, acknowledging his loss of control, abandoned us and disappeared behind a screen.

In my mind I could hear, "Do something, Mother. Do something!" So I did.

I bent over, disentangled the gibbons, set the microphone back up on its round base and began my discourse with: "These are not monkeys—they are apes!" I then intended to tell the crowd all about gibbons and maybe I did, but when something that didn't sound at all like my voice shouted the words back at me, so great was my terror and so difficult my job of keeping Penny and Pogo from swinging on the microphone again, that I have no recollection of what I said—if anything.

I must have told them very little because the question period went on and on. I think I remember saying, when asked what Penny and Pogo ate, "Please do not give them anything with sugar. It makes them very sick." If I said it—it bore no fruit.

Finally, rescued by the emcee, we left the stage. The crowd followed us to the car handing the gibbons cookies, cake, grape pop and even an ice cream cone. Ampon trotted behind, snatching away what contraband she could and offering the gibbons the grasshoppers instead, but Penny and Pogo were smart and clung to the goodies. What they couldn't cram into their mouths, they spilled on me.

The sugar made its rapid way through the gibbons before we got home. The cut-down diapers were too flimsy and too few for the volume. The car was a mess. Fortunately, unlike Pruno's calamity on the day we left for Bangkok, the mess wasn't very odoriferous; but it did entail a thorough washing out of the station wagon.

After the dispensation of the Pepto Bismol, when Penny and Pogo were in bed, as I soaped and mopped the car's leather seats and the plush carpeting, I thought about Pruno and that awful exodus from Vienna with a certain nostalgia. I also thought about the day after, and began to laugh at the memory of my conversation with mother's cook, Ollie, as I had sloshed and deodorized that other station wagon four years before.

Ollie had sat on the back steps and had kept me entertained as I worked.

Her malapropisms and her way of telling things were unforgettable and I could again hear her perfect descriptions uttered in her matter-of-fact tone.

"Dinky had the diree once," Ollie had mused. Dinky was the youngest of the thirteen children. There were four sets of twins and five singletons. "Yes, ma'am," she'd continued when I'd looked up at her, "he shor was sick."

"Poor little thing," I'd commiserated. "What did you do for him?"

"Well, Robert's mama (Robert was her husband) said to give him some Epsom Sauce, so we done that."

My back had stiffened and I'm sure my eyes had grown wide in disbelief. "How old was he then, Ollie?" He was only five at the time of this recitation.

"Oh, he was right little. Weren't no bigger than a sparrow. Must have been five, maybe six months old. Silly ole woman, Robert's mama. Doctor, when he come, said we might ought to have killed Dinky. Dr. Newman, he took us right to the hospital in his very own car."

"What made her think that Epsom Salts would be all right for a baby?" I'd asked.

"She said it would rench off his blood. Dang near renched it right out of his body. He really had the diree then. Dr. Newman was real mad at that ole woman."

I told her I hoped she never listened to her mother-in-law again.

"Tried not to," she answered, "but she can be kinda mean when she don't get paid no mind to."

I waited. My silence was rewarded.

"One day, guess he was maybe three year old, he took to grindin' his teeth. Awful soundin' it was. Made my teeth shiver. Mama Flowerine says, 'The kid's got worms.' Went right down to Hawthorne's Drug Store she did and bought some pills. They was for dogs, but she says Dinky was dog size and they'd work just fine. Give 'em to him and the next morning he had the measles. Don't reckon the pills done it cause Robby had the measles too and he hadn't had no pills, but Dinky was a lot sicker on account of he got the diree again."

As I finished cleansing the second car of a diarrhea disaster, I remembered that Dinky was now nine and seemed likely to make it to adulthood and freedom from Mama Flowerine's ministrations. From the bottom of my heart I hoped so, and even more, I hoped I wouldn't have to perform this labor ever again!

XIX

The phone rang! "Hello," I said.

"Is this the Vanderhoef residence?" A pleasant, well-modulated voice waited for my answer.

"Yes, it is. This is Mrs. Vanderhoef speaking."

"Just the person I wanted," the voice continued. "This is Miss Patterson from Parade Magazine. Do you know what Parade Magazine is?"

"Yes, of course I do. We take the Washington Post." I wondered what was coming.

"We understand you have a pair of gibbons." There was a slight hesitation. "Could we come out to see them and to interview you?"

Parade Magazine! Our fame was spreading. So far we were well known only in our own area. "I'd be delighted to have you come out. What day would you like to come?"

"How about tomorrow? I know it's a little soon, but we'd like to do the interview as soon as possible. Would tomorrow be convenient for you?"

"Tomorrow would be fine," I said. "The weather is supposed to be good and the gibbons are at their best outdoors." This was late spring of their first year in Virginia.

"May I bring a photographer with me?" Miss Patterson, I'm sure, already knew the answer, but she had to ask.

"Of course." We arranged the time, I gave her directions to Meadowlark Road, explaining the necessity of putting her radio antenna down before she drove in the driveway, and we hung up. We had a whole day for the family to get excited. A national newspaper magazine! We'd be on everyone's Sunday breakfast table.

Ampon and her husband, Sampow, who'd arrived in March, flew to polish the already spotless house and to trim the already manicured lawn. Even Christy picked up the trash in her room. That she stuffed it all into the closet and under her bed made no difference. Her sunny yellow room looked almost clean.

The interview day was Saturday. Van went off to NSA in Maryland (the army does not always recognize holidays) and Craig went to his after-school and Saturday job at McPhail's Hardware Store. The rest of us waited.

At eleven o'clock a car stopped at the end of the drive, a young man emerged and, according to my instructions, put the radio antenna down. Parade Magazine was arriving.

The gibbons rushed to greet the strangers who, at first, were a little cautious, but soon everyone was a friend. Miss Patterson, Julia, was attractive and bouncy, and she loved the gibbons. Pogo thought she was there just to see him, and after we had settled down on the comfortable wicker furniture under the trees, one by one he brought her the toys he had scattered around the lawn.

The photographer was an opinionated young man who tried to impress us with his vast knowledge, but whose dealings with gibbons were not part of his expertise. He antagonized Penny by paying no attention at all to her advances, and Pogo became irritated at the man's lack of generosity when he wouldn't share his iced tea. Both gibbons seemed to dismiss him from the crowd and, we thought, ignored him.

When, at Julia's suggestion, the photographer finally got out his camera, he filled the pockets of his sport coat with filters from the roomy black leather bag he carried over his shoulder, and then hung the jacket on a chair back.

"I wouldn't do that if I were you," I said to him in warning.

"Why not?"

"You won't have your filters long."

He shrugged his shoulders and looked at me in pity for my stupidity. "The monkeys are in the trees. They'll never know the filters are there." His use of the word "monkeys" alienated me immediately. I knew the filters were expensive and I was torn between that fact and the desire to have the young man taught a lesson. My better self won. I tried again and was rebuffed. All right, I thought, learn it the hard way.

While Penny swung down from the tree and onto the table to dip her hand in the photographer's unprotected iced tea and to distract him, Pogo grabbed a filter from its hiding place in a pocket and went up the tree. I watched as he positioned himself very carefully above Hugh, the young man. I heard a small giggle and realized that Julia was watching, too.

The big, round, heavy, amber filter dropped. It hit Hugh's shoulder with an awful thwack and rolled under the table.

As the young man clutched his shoulder, the rest of us bent to retrieve the filter. Peering under the table, I could see the jacket hanging not far from my face. I watched silently as both Penny and Pogo reached into the pockets to grab handfuls of smaller filters. Juggling expertly, the gibbons climbed to the lower branches of a tulip poplar.

When the round, hard objects began to rain from above, Hugh became apoplectic. He was almost as mad at the rest of us as he was at the gibbons, because we were amused, but by the time he'd collected all the unbroken filters, he'd collected himself as well and was almost enjoying the joke. I liked him better. He took some wonderful pictures of gibbons swinging in the trees, gibbons stealing iced tea, gibbons riding the goats, gibbons eating their lunch, gibbons being affectionate and, in general, a day in the life of a Vanderhoef gibbon.

After the excellent article came out, letters came from all over the east coast. Most writers wanted to know more about the gibbons, but some were from old friends with whom we had lost touch; we were delighted to find them again.

After a year or so, a problem began to develop in our arrangements. Penny became unpredictable with little boys. Basically, she developed an antipathy to them so that when Dang and his little friends were playing at our house, we had to make sure they were nowhere in Penny's vicinity or someone would be bitten. She was beginning to get her second teeth, two of which were long, sharp eye-teeth, and she could now inflict some damage. We watched carefully and there were no major disasters although there were a few, what might be called, nips.

When Penny's aversions spread to little girls and then to all men except Van and Craton Guthrie, we strung a wire between two trees. The trees were almost forty feet apart and the chains that attached to Penny and Pogo's belts were long enough so they could climb into the lower branches. Gibbons and visitors were safe.

Penny attached to the wire.

In the afternoons on nice days, I took the gibbons back into the woods and sat under the trees with a book, while they, free to swing and behave like gibbons, explored the two hundred year old tulip poplars and hickories. I had to go alone on these excursions or with a female friend, as Penny would no longer tolerate Van or Craig. In the winter, we still spent our afternoons in the family room.

One afternoon, in the fall of 1961, Lee and I brought the gibbons in from outside in time for their supper. They'd lived with us for five years by then and the routine was well established. Lee was taking off Pogo's belt and apparently caught a little of his fur. He squeaked; Penny tore away from me and before I even thought to grab her, she attacked Lee and tore Lee's ear lobe. Fortunately, it healed well, but I knew then that something had to be done before we had a greater tragedy.

I called Dr. Theodore Reed, the director of the National Zoo in Washington. I'd been in contact with Dr. Reed before, and I wanted to ask about having Penny and Pogo's eye teeth removed. Gibbons couldn't withstand anesthetic, I was told.

"Please, Mrs. Vanderhoef, find a zoo that will accept them while you still have a face without scars," the zoo director advised. "Gibbons are wild animals and, as such, are subject to what we call 'red-outs.' Now that Penny has reached adolescence and Pogo is almost there, if they become upset or frightened, their little brains will explode in frenzy and—for long enough to do considerable damage—they won't know what they're doing. Penny will be terribly contrite when her mind clears, but the damage will have been done."

"Is there no other alternative?" I asked, my heart the weight of a bowling ball in my chest.

"I can offer you none." Dr. Reed understood how I felt and his voice was filled with sympathy.

"Could the National Zoo take them? That way I would be able to visit them."

"Mrs. Vanderhoef, we have an already established colony of gibbons that has grown too large. We have two yearlings we are trying to find homes for, so, I'm sorry, but taking yours is out of the question. Besides, I'm sure you know (I didn't know until that moment) that you can't put strange gibbons into an already established colony. They would be torn to pieces immediately by the colony residents."

"Where can I find a zoo without a colony?"

"I am probably cutting the throat of my own zoo," the concerned man said, "but you might try the Cleveland Zoo. They don't as yet, have any gibbons, and Dr. Goss was saying, just the other day, that their new gibbon room is almost ready. Let me give you Dr. Goss's phone number and, if I were you, I'd call him right away before someone else, namely the National Zoo, beats you to it."

"I do thank you," I said, hoping the tears streaming down my face didn't show in my voice. "I will call them right away before I lose my determination and just go off to live in the woods alone with my gibbons."

"I understand your pain, but believe me, they have each other and will soon be at the mating age when you will be superfluous anyway. Do make the arrangements, so that, like our human children, they can have an independent life of their own."

I thanked him again and hung up the phone. As soon as my tears stopped, I called Dr. Goss at the Cleveland Zoo.

"Dr. Goss," I said, "this is Mrs. Vanderhoef in Vienna, Virginia. Dr. Reed at the National Zoo suggested I call you. . . ." My voice became a gurgle.

"Take your time, Mrs. Vanderhoef." The compassionate person sitting someplace on the shore of Lake Erie in Cleveland, understood my problem. He'd probably met it before. "When you can, tell me what kind of animal you want to talk about."

"A pair of gibbons." My voice came out in a small squeak.

"How old are they? Take your time. Don't answer till you're able."

"I think I'm under control now." I drew a deep breath, swallowed and said, "The girl is six and the boy is about six months younger. Dr. Reed says they will soon be at the mating age."

Dr. Goss kept me talking about the gibbons while at the same time he let me know what kind of facility they had in Cleveland, and how our precious babies would be cared for. As Dr. Reed had, he told me that both gibbons would really be capable of seriously damaging somebody, and he hoped we wouldn't delay too long in making our decision. As for his decision, since we were donating the gibbons, he didn't even have to ask the board of directors of the zoo, and, yes, they would be happy to have Penny and Pogo as the founders of their gibbon colony.

"I'll call you when I've made their travel arrangements," I said. "Our daughter is having knee surgery next week and, if I can arrange it, Penny and Pogo will be shipped the same day so our minds will be occupied. I won't back down, Dr. Goss. Penny and Pogo will come to live with you." I was conscious of the weight of my heart underneath my ribs, and tears plopped onto, and stained, my dark green shirt.

I went to tell Ampon, who loved the gibbons as much as I did. We cried together the rest of the afternoon. Dinner was dismal. Everyone wept even though none of the other family members loved the gibbons as Ampon and I did, and certainly Dang, who'd been nipped more than once, had no reason to care about them at all.

After dinner I called Craton Guthrie because of his love for our little apes, particularly Penny. He, too, was devastated, but . . . "I'm flying the Cleveland run next week," he said. My heart lifted. They would at least go with someone they knew and cared about, and Craton would make sure they got from the airport to the zoo safely.

"I'll take care of their reservations tomorrow." Craton took that responsibility from me. "And, if possible, I'll put their crate right up behind me in the cockpit. Without question I'll go with them to the zoo and if I don't like their accommodations, I'll bring them right back home. How's that for service?"

"Craton, man or beast, no one ever had a better friend."

"Thank you, Jeannie. I'll call you tomorrow when the arrangements are made." We hung up.

Lee had torn the ligaments and cartilage in her knee riding a horse other than her own, and her surgery was scheduled for Wednesday morning.

Tuesday afternoon we delivered her to the hospital, went through the bloodwork and spoke with the anesthesiologist. Ampon was going to take Penny and Pogo for one last romp in the woods and my heart was torn in two, wanting to be with Lee and, at the same time, wanting to be with my beloved furry children on their last afternoon at home.

When Van and I got home, Ampon, both legs in bloody bandages, was hobbling around the kitchen fixing dinner.

"What happened?" we asked as we led her to the breakfast table and put her into a chair.

"We go woods, come back, I give chan-nee supper, put they in room. Somesing make Penny angry and she bite me bad in my legs! Not do nossing to she. She just run across floor and bite!"

"We'd better get to the hospital. Come, get your coat."

"Madame Guthrie she took me to see Dr. Newman. He put medicine on and say change bandages plitty soon."

"Here, I'll change them now, and then you go rest in bed. I'll finish the dinner."

Ampon argued, but I won. When I uncovered the bites I saw that they were actually tears from the long eye-teeth, some as long as two inches. They weren't terribly deep, but they would leave scars and they were painful. Thank heaven the decision had already been made and the zoo was waiting.

The next morning when Ampon, determined to hold the gibbons one more time, appeared at their door, Penny flew toward her in a terrible rage and Ampon couldn't go in.

I opened the door in fear, but Penny and Pogo were as loving and affectionate with me as usual. We had a quiet hour together, I put them in their room and closed the door for the last time.

Craton, when he came, would put them in their crate and would take them with him all the way to the Cleveland Zoo.

Most of my tears had been shed in the last week, and when Van and I drove off to the hospital to be with Lee, I took a deep breath and jostled the heaviness in my heart so that the worry about my daughter was uppermost.

Lee's operation went fine and she rather enjoyed being waited on hand and foot in the Northern Virginia Doctor's Hospital.

When Craton got home from Cleveland he reported that the gibbon room in Cleveland was enormous. It was glass-enclosed, Penny and Pogo had their own kitchen and their own keeper, and they were right next to Dr. Goss's office. If not ecstatically happy, they were, at least, content.

Many times in the months that followed I missed those furry arms around my neck or the little hands that slid into mine as I walked across the yard. Often I started down to their room to get them before I remembered they weren't there. Often I longed to have a furry hand grab my sandwich or pull my hair and, most of all, to have two pansy shaped faces grinning at me in mischief or in love. I will always miss them! Still, I give thanks from the bottom of my heart, that for a while, at least, Penny and Pogo had shared a part of my life!

XX

"Hello, Dr. Goss? This is Jeanne Ann Vanderhoef in Vienna, Virginia."

"Oh, yes, Mrs. Vanderhoef. How are you? We haven't heard from you for quite a while and were wondering, just the other day, if everything was all right with the Vanderhoef family."

"Oh, we're fine. I had just thought it was time to give you a rest from my inquiries. I know Penny and Pogo are in good shape; otherwise, you'd have given us a call. What I'm calling about today is something quite different."

"You don't have any more animals to give us, do you? I hope not, because we're just about filled up."

"No." I laughed. "We're going to be driving through Cleveland and would like to stop to see you if possible. I do hope you won't be off on one of your 'beast trading' trips."

"I know you're really coming to see Penny and Pogo and it doesn't hurt my feelings one bit. We'll be happy to have you any time. We, also, are looking forward to meeting you. I'll be in Washington in three months for the annual 'animal swap' and was going to call you then. This is better. When are you coming?"

"We'll be there Wednesday evening and would like to

come to the zoo as early Thursday morning as possible. What time do you open?

"Nine o'clock, but if you'd like to come earlier . . . ."

"No, that's fine . . . and Dr. Goss . . . I don't want Penny and Pogo to see us; it will just upset them. Is there some way we can be hidden and still watch them? I just want to know they're all right and happy."

"Oh, don't worry. They've been here a year and a half. They won't even remember you. I hope that doesn't hurt you, but that's the way animals are."

Not our animals, I thought. "I'll call you Wednesday when we get to the motel," I said and we hung up.

Van and I were on our way to Van's home in North Dakota to see Craig graduate from the University there. It was a joyous occasion, of course, but for me the occasion was colored by the eagerness to see Penny and Pogo, and the dread of leaving them again. I hoped I could get through it with some propriety, but wasn't confident. "I'll just peek at them enough to know they're happy," I told myself and Van, "and then we'll leave." Van said afterwards he knew I was whistling in the dark.

We got to the motel in Cleveland on schedule, I called Dr. Goss on schedule, and we were at the zoo gates on Thursday morning on schedule. From then on, the day fell apart.

Dr. Goss was a charming man in his early fifties. As he led us along the paths toward the building that housed his office and Penny and Pogo's home, excitement fluttered and bubbled in my heart causing it to alternately race, or to stop beating all together. I hardly noticed the animals we passed or the fact that, as early as it was, there was already a crowd of children and parents. Most schools were just out for the summer and it was a gorgeous day. Families were taking advantage of the ideal temperature.

We filed into the huge building behind a mob of assorted ages and heights. Inside, a wide aisle ran between large, well-lighted, glass-enclosed rooms, each holding its

Penny and Pogo at the Cleveland Zoo.

own pair of animals. These were the creatures that would be transferred outside when the weather was warm enough but who lived indoors in the winter months.

In one room, seemingly quite content, a pair of leopards rough-housed on the spotlessly clean floor. In the room next to them, two baby chimpanzees chased each other around, and over, their jungle gym. The walls dividing the rooms were not glass and no occupant could see his neighbors, so the chimps, for instance, were in no fear of the leopards.

Instinct is strong and I remembered the gibbons' terror of a toy leopard someone once brought to Christy. That had given us a clue to the gibbons' reaction when a black and white spotted Great Dane had run through our yard in Vienna and Penny and Pogo had filled the woods with loud "whoops" of terror and warning. They had thought Certes was a leopard. Certes, unfortunately, died of a snake bite before the gibbons learned to accept him as they had accepted tiger-like Pruno.

The room next to the chimpanzees had a large crowd in front of it and people were laughing. "Oh," I said to Dr. Goss, "that has to be the gibbons' room and Pogo must be performing."

"He does enjoy an audience and he does keep them amused." Dr. Goss smiled. "Penny and Pogo are everyone's favorites."

We had reached the edge of the crowd and as I slid in behind a man holding a little boy on his shoulders, I saw Pogo grab an orange from the floor and race up the jungle gym to sit on top and to assess the assembly outside his glass stage. Penny was swinging by an arm from the same contrivance enjoying a banana. When Pogo's bright eyes scanning the crowd, met mine, the orange dropped.

"Eeeee!" He shrieked. "Eeeee! Eeeee!" He swung down and ran to the window. Penny looked out, saw me, threw away the banana and rushed with Pogo across the floor. I had the feeling that for the past year and a half they had known this moment would come.

Drawn, as if by magnet, I pushed my way to the

window. We met, the glass a barrier between my yearning arms and the furry ones so anxious to twine around my neck.

I pushed so hard trying to reach through, my hands on the glass must have left indelible prints. Penny and Pogo's tiny hands pushed equally as hard from the other side. Hearts broke, but the thick, plate glass didn't.

I felt Van's arm around my shoulder trying gently to move me away. I couldn't leave! Tears flowed unchecked like the Mississippi. The crowd was silently watching what I'm sure even the most insensitive realized was a tragic moment. I didn't care!

"Come in the back." Dr. Goss took my arm. "You can see their sleeping compartment and feeding room and you can touch them there." I knew there was a chance Van would continue the trip alone if I once touched them, but I let myself be led. The longing was too strong!

The top half of the door that opened from the kitchen shared by the chimps and the gibbons was half-glass. Penny hung on the door, so sure that I would come in and take her in my arms. She alternately cried and grinned as she hung from the door frame and pounded her feet against the glass. Pogo came into the feeding compartment where we could touch each other. At the feel of his furry head and warm little hand, my already shattered heart began to lose some of its pieces. I felt them drop away into some sort of a dark void.

"Would you like to go in with them?" Dr. Goss's voice was compassionate.

"Dr. Goss," I said, the Mississippi still flowing, "if I take them in my arms I will either spend the rest of their lives with them in your zoo, or I will grab them and run away!" I knew it was true and it was time to go. In my selfishness I had made a tragic mistake in coming here and I knew it. I couldn't give any of us the thing we wanted most, and again . . . I had to betray my precious, trusting babies. My eyes still on them, I backed away until, as I rounded a corner, we lost sight of each other forever.

I thought perhaps my heart no longer had a bottom. I thought perhaps it would always remain open and sore and that anything that went into it in the way of love would immediately drain away. That of course, is not so. My heart has healed and the love is still in it, but so is the pain when I re-live that terrible time.

From the very bottom of my heart . . . today . . . my personal Mississippi is flooding.

Epilogue

A month or so after we came home from our terrible day in Cleveland, I wakened early one morning quite sure that I had heard Penny whooping in the woods behind the house. In the kitchen, Ampon had the back door open and was listening to something I couldn't hear.

"I sink I hear Penny!" she said when I asked her what she was doing.

"Oh, Ampon! I thought I did too. That's what wakened me. Maybe I'd better call the zoo."

"Madame, wait. If somesing wrong wis Penny, man will call."

All the same, Ampon marked the date, July eleven, on the calendar. No one called.

That evening we had friends in for cocktails. We were sitting in the living room, the conversation friendly and flowing, when suddenly Van said loudly, "Would you look at that!" He pointed to the large window in one end of the living room.

Naturally, everyone looked. There, on the outside of a window that had been washed innumerable times in the almost two years she had been gone ... were Penny's footprints.

She had been there, trying to get in. It took a lot of persuading to keep me from running into the woods to look for her, but in my heart I knew she wasn't there.

When Dr. Goss came to Washington for the animal "swap" three months later, he called.

"I didn't want to just write a cold letter," he said, "but Penny never gave up trying to get out after your visit. Our Penny is gone."

My heart sank. Penny escaped, I thought, and is some place in the wide world cold and alone. Maybe Ampon and I did hear her in the woods. "What do you mean, gone? Did she escape?"

"No," Dr. Goss's voice faltered. "She died of pneumonia. After you left she flew off the trapeze time after time and hit the half-inch plate glass with her feet until she finally cracked it. It took her three weeks to manage this, but she kept at it. We replaced it with three-quarter inch glass which she never succeeded in breaking, though she spent her days trying. When she got pneumonia she made no effort to fight it and was gone in eighteen hours."

Again, that heavy, bulky lump slid down from my throat to lodge in the bottom of my heart.

"What was the date, Dr. Goss?" I managed to ask.

He gave me the date, July eleven. I told him about the early morning of July eleven and the marked calendar. "She made it home," I managed to say through my tears, "and she left us her footprints. Ampon won't wash them away and I don't want her to."

"Mrs. Vanderhoef." The zoo director's voice was very quiet. "Penny was pregnant when she died. You would have been a grandmother."

"What about Pogo?" I asked at last. "Is he all right?"

"He's just fine." Dr. Goss sounded relieved that the conversation had shifted. "We got him a new little wife, but he would have none of her at first. Now, though, they seem to be friends. She's still too young to mate, however."

"I'm glad he's not alone, but he's so happy-go-lucky he won't be lonely long. Dr. Goss, Penny was dependent on me, Pogo was more dependent on her. He won't miss us as she did so I probably won't call you again. If you need us for any reason, please let us know!"

We have heard no more. Although Pogo must have died by now, I don't know when. I do know that, although the footprints have gone from the window and our gibbons know what the "hereafter" is, those indelible prints are still in my heart!

☙ ☙ ☙